Intelligent Systems Reference Library

Volume 129

Series editors

Janusz Kacprzyk, Polish Academy of Sciences, Warsaw, Poland
e-mail: kacprzyk@ibspan.waw.pl

Lakhmi C. Jain, University of Canberra, Canberra, Australia;
Bournemouth University, UK;
KES International, UK
e-mail: jainlc2002@yahoo.co.uk; jainlakhmi@gmail.com
URL: http://www.kesinternational.org/organisation.php

About this Series

The aim of this series is to publish a Reference Library, including novel advances and developments in all aspects of Intelligent Systems in an easily accessible and well structured form. The series includes reference works, handbooks, compendia, textbooks, well-structured monographs, dictionaries, and encyclopedias. It contains well integrated knowledge and current information in the field of Intelligent Systems. The series covers the theory, applications, and design methods of Intelligent Systems. Virtually all disciplines such as engineering, computer science, avionics, business, e-commerce, environment, healthcare, physics and life science are included.

More information about this series at http://www.springer.com/series/8578

Margarita-Arimatea Díaz-Cortés
Erik Cuevas · Raúl Rojas

Engineering Applications
of Soft Computing

Springer

Margarita-Arimatea Díaz-Cortés
Department of Mathematics and Computer
  Science
Freie Universität Berlin
Berlin
Germany

Raúl Rojas
Department of Mathematics and Computer
  Science
Freie Universität Berlin
Berlin
Germany

Erik Cuevas
CUCEI, Universidad de Guadalajara
Guadalajara, Jalisco
Mexico

ISSN 1868-4394   ISSN 1868-4408   (electronic)
Intelligent Systems Reference Library
ISBN 978-3-319-86257-6   ISBN 978-3-319-57813-2   (eBook)
DOI 10.1007/978-3-319-57813-2

Printed on acid-free paper

This Springer imprint is published by Springer Nature
The registered company is Springer International Publishing AG
The registered company address is: Gewerbestrasse 11, 6330 Cham, Switzerland

Preface

Currently, we are faced with many complex engineering systems that need to be manipulated. As they are usually not fully theoretically tractable, it is not possible to use traditional deterministic methods. Soft computing, as opposed to conventional deterministic methods, is a set of methodologies (working synergistically, not competitively) that, in one form or another, exploits the tolerance for imprecision, uncertainty and approximate reasoning to achieve tractability, robustness, low-cost solution, and close resemblance with human-like decision making. Soft computing techniques include neural networks, evolutionary computation, fuzzy logic and Learning Automata. The recent years have witnessed tremendous success of these powerful methods in virtually all areas of science and technology, as evidenced by the large numbers of research results published in a variety of journals, conferences, as well as many books.

Engineering is a rich source of problems where each new approach that is developed by mathematicians and computer scientists is quickly identified, understood and assimilated in order to be applied to specific problems. In this book we strive to bring some state of the art techniques by using recent Soft Computing techniques after its application to challenging and significant problems in engineering.

Soft Computing methods have many variants. There exist a rich amount of literature on the subject, including textbooks, tutorials, and journal papers that cover in detail practically every aspect of the field. The great amount of information available makes it difficult for no specialist to explore the literature and to find the right technique for a specific engineering application. Therefore, any attempt to present the whole area of Soft Computing in detail would be a daunting task, probably doomed to failure. This task would be even more difficult if the goal is to understand the applications of Soft Computing techniques in the context of engineering application. For this reason, the best practice is to consider only a representative set of Soft Computing approaches, just as it has been done in this book.

The aim of this book is to provide an overview of the different aspects of Soft Computing methods in order to enable the reader in reaching a global understanding of the field and, in conducting studies on specific Soft Computing techniques that are related to applications in engineering, which attract the interest for their complexity. Our goal is to bridge the gap between Soft Computing techniques and their applications to complex engineering problems. To do this, at each chapter we endeavor to explain basic ideas of the proposed applications in ways that can be understood by readers who may not possess the necessary backgrounds on some of the fields. Therefore, engineers or practitioners who are not familiar with Soft Computing methods will appreciate that the techniques discussed are beyond simple theoretical tools since they have been adapted to solve significant problems that commonly arise on such areas. On the other hand, members of the Soft Computing community can learn the way in which engineering problems are solved and handled by using intelligent approaches.

This monograph presents new applications and implementations of Soft Computing approaches in different engineering problems. The present book collects 12 chapters. It has been structured so that each chapter can be read independently from the others. Chapter 1 describes the main methods that integrate Soft Computing. This chapter concentrates on elementary concepts of intelligent approaches and optimization. Readers that are familiar with such concepts may wish to skip this chapter.

In Chap. 2, a Block matching (BM) algorithm that combines an evolutionary algorithm (such Harmony Search) with a fitness approximation model is presented. The approach uses motion vectors belonging to the search window as potential solutions. A fitness function evaluates the matching quality of each motion vector candidate. In order to save computational time, the approach incorporates a fitness calculation strategy to decide which motion vectors can be only estimated or actually evaluated. Guided by the values of such fitness calculation strategy, the set of motion vectors is evolved through evolutionary operators until the best possible motion vector is identified. The presented method is also compared to other BM algorithms in terms of velocity and coding quality. Experimental results demonstrate that the presented algorithm exhibits the best balance between coding efficiency and computational complexity.

Chapter 3 presents an algorithm for the optimal parameter identification of induction motors. To determine the parameters, the presented method uses a recent evolutionary method called the Gravitational Search Algorithm (GSA). Different to the most of existent evolutionary algorithms, GSA presents a better performance in multimodal problems, avoiding critical flaws such as the premature convergence to sub-optimal solutions. Numerical simulations have been conducted on several models to show the effectiveness of the proposed scheme.

In Chap. 4, an image segmentator algorithm based on Learning Vector Quantization (LVQ) networks is presented and tested on a tracking application. In LVQ networks, neighboring neurons learn to recognize neighboring sections of the input space. Neighboring neurons would correspond to object regions illuminated in a different manner. The segmentator involves a LVQ network that operates directly on the image pixels and a decision function. This algorithm has been applied to color tracking, and have shown more robustness on illumination changes than other standard algorithms.

Chapter 5 presents an enhanced evolutionary approach known as Electromagnetism-Like (EMO) algorithm. The improvement considers the Opposition-Based Learning (OBL) approach to accelerate the global convergence speed. OBL is a machine intelligence strategy which considers the current candidate solution and its opposite value at the same time, achieving a faster exploration of the search space. The presented method significantly reduces the required computational effort yet avoiding any detriment to the good search capabilities of the original EMO algorithm. Experiments are conducted over a comprehensive set of benchmark functions, showing that the presented method obtains promising performance for most of the discussed test problems.

In Chap. 6, the use of the Learning Automata (LA) algorithm to compute threshold points for image segmentation is explored. Despite other techniques commonly seek through the parameter map, LA explores in the probability space providing better convergence properties and robustness. In the chapter, the segmentation task is therefore considered as an optimization problem and the LA is used to generate the image multi-threshold separation. In the approach, a 1D histogram of a given image is approximated through a Gaussian mixture model whose parameters are calculated using the LA algorithm. Each Gaussian function approximating the histogram represents a pixel class and therefore a threshold point. The method shows fast convergence avoiding the typical sensitivity to initial conditions such as the Expectation-Maximization (EM) algorithm or the complex time-consuming computations commonly found in gradient methods. Experimental results demonstrate the algorithm's ability to perform automatic multi-threshold selection and show interesting advantages as it is compared to other algorithms solving the same task.

Chapter 7 describes the use of an adaptive network-based fuzzy inference system (ANFIS) model to reduce the delay effects in gaze control and also explains how the delay problem is resolved through prediction of the target movement using a neurofuzzy approach. The approach has been successfully tested in the vision system of a humanoid robot. The predictions improve the velocity and accuracy of object tracking.

Chapter 8 presents an algorithm for the automatic detection of circular shapes from complicated and noisy images with no consideration of the conventional Hough transform principles. The presented algorithm is based on an Artificial Immune Optimization (AIO) technique, known as the Clonal Selection algorithm (CSA). The CSA is an effective method for searching and optimizing following the Clonal Selection Principle (CSP) in the human immune system which generates a response according to the relationship between antigens (Ag), i.e. patterns to be recognized and antibodies (Ab) i.e. possible solutions. The algorithm uses the encoding of three points as candidate circles over the edge image. An objective function evaluates if such candidate circles (Ab) are actually present in the edge image (Ag). Guided by the values of this objective function, the set of encoded candidate circles are evolved using the CSA so that they can fit to the actual circles on the edge map of the image. Experimental results over several synthetic as well as natural images with varying range of complexity validate the efficiency of the presented technique with regard to accuracy, speed, and robustness.

In Chap. 9, an algorithm for detecting patterns in images is presented. The approach is based on an evolutionary algorithm known as the States of Matter. Under the proposed method can be strongly reduced the number of search locations in the detection process. In the presented approach, individuals emulate molecules that experiment state transitions which represent different exploration–exploitation levels. In the algorithm, the computation of search locations is drastically reduced by incorporating a fitness calculation strategy which indicates when it is feasible to calculate or only estimate the Normalized cross-correlation values for new search locations. Conducted simulations show that the presented method achieves the best balance over other detecting algorithms, in terms of estimation accuracy and computational cost.

Chapter 10 explores the use of the Artificial Bee Colony (ABC) algorithm to compute threshold selection for image segmentation. ABC is a heuristic algorithm motivated by the intelligent behavior of honey-bees which has been successfully employed to solve complex optimization problems. In this approach, an image 1D histogram is approximated through a Gaussian mixture model whose parameters are calculated by the ABC algorithm. For the approximation scheme, each Gaussian function represents a pixel class and therefore a threshold. Unlike the Expectation-Maximization (EM) algorithm, the ABC-based method shows fast convergence and low sensitivity to initial conditions. Remarkably, it also improves complex time-consuming computations commonly required by gradient-based methods. Experimental results demonstrate the algorithm's ability to perform automatic multi-threshold selection yet showing interesting advantages by comparison to other well-known algorithms.

In Chap. 11, the usefulness of planning to improve the performance of feedback-based control schemes considering a probabilistic approach known as Learning Automata (LA) is considered. Standard gradient methods develop a plan evaluation scheme whose solution lies on a neighborhood distance from the previous point, forcing to explore the space extensively. On the other hand, LA algorithms are based on stochastic principles, with newer points for optimization being determined by a probability function with no constraint whatsoever on how close they are from previous optimization points. The presented LA approach may be considered as a planning system that chooses the plan with the higher probability to produce the best closed-loop results. The effectiveness of the methodology is tested over a nonlinear plant and compared with the results offered by the Levenberg-Marquardt (LM) algorithm.

Finally, Chap. 12 presents an algorithm based on fuzzy reasoning to detect corners even under imprecise information. The robustness of the presented algorithm is compared to well-known conventional corner detectors and its performance is then tested on a number of benchmark images to illustrate the efficiency of the algorithm under uncertainty conditions.

As authors, we wish to thank many people who were somehow involved in the writing process of this book. We express our gratitude to Prof. Lakhmi, who so warmly sustained this project. Acknowledgements also go to Dr. Thomas Ditzinger, who so kindly agreed to its appearance. We also acknowledge the support of the Heinrich Böll foundation and CONACYT, under the grant numbers P124573 and CB 181053 respectively.

Berlin, Germany   Margarita-Arimatea Díaz-Cortés
March 2017   Erik Cuevas
                                                                                      Raúl Rojas

Contents

Chapter 1
Introduction

This chapter gives a conceptual overview of the soft computing techniques and optimization approaches, describing their main characteristics. The goal of this introduction is to motivate the consideration of soft computing methods for solving complex problems. On the other hand, the study of the optimization methods is conducted in such a way that it is clear the necessity of using intelligent optimization methods for the solution of several engineering problems.

1.1 Soft Computing

Soft computing (SC) [1] is a computer science area that tries to develop intelligent systems. Under soft computing, the objective is to produce computer elements which artificially operate based on intelligent processes typically extracted from natural or biological phenomena.

SC considers a set of methodologies that try to develop systems with tolerance to the imprecision and robustness to the uncertainty. SC techniques have demonstrated their capacities in the solution of several engineering applications. Different to classical methods, soft computing approaches emulate cognition elements in many important aspects: they can acquire knowledge from experience. Furthermore, SC techniques employ flexible operators that perform input/output mappings without the consideration of deterministic models. With such operators, it is possible to extend the use of SC techniques even to those applications in which the precise and accurate representations are not available.

SC involves a set of approaches extracted from many fields of the computational intelligence. SC considers three main elements [2]: (A) fuzzy Systems, (B) neural networks and (C) evolutionary computation. Furthermore, other alternative approaches have been associated with SC. They include techniques such as machine learning (ML) and probabilistic reasoning (PR), belief networks, and expert systems, etc. Figure 1.1 shows a conceptual map that describe the different branches of SC.

© Springer International Publishing AG 2017
M.-A. Díaz-Cortés et al., *Engineering Applications of Soft Computing*,
Intelligent Systems Reference Library 129, DOI 10.1007/978-3-319-57813-2_1

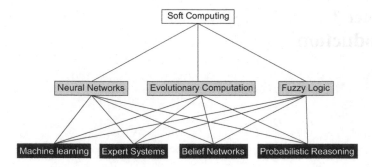

Fig. 1.1 Different branches of soft computing

Soft-computing, in general, considers the inspiration from natural phenomena or biological systems. Neural networks emulate the animal brain connections; evolutionary algorithms base its operation on the characteristics of Darwinian evolution. Meanwhile fuzzy logic aims to simulate the highly imprecise nature of human reasoning.

1.2 Fuzzy Logic

Humans handle imprecise and uncertain information in their day life routines. This fact can be observed in the structure of the language which includes various qualitative and subjective terms and expressions such as "quite expensive", "very old", or "pretty close", "cheap", etc. In human knowledge, approximate reasoning is employed and considered to provide several levels of imprecision and uncertainty in the concepts that are transmitted among persons.

Fuzzy logic [3] represents a generalization of classical logic. It considers fuzzy sets which are an extension of crisp sets in classical logic theory. In classical logic, an element belongs to either a member of the set or not at all, in fuzzy logic, an element could belong at a certain level "degree of membership" to the set. Hence, in fuzzy logic, the membership function of an element varies in the range from 0–1, but in classical logic, the membership value is only 0 or 1.

A fuzzy system involves four different modules: a fuzzy rule base, a fuzzification module, an inference engine, and a defuzzification module. Such elements are shown in Fig. 1.2. The fuzzification module translates the input values into fuzzy values which are used by the fuzzy system. The inference engine considers the outputs of the fuzzification element and uses the fuzzy rules in the fuzzy rule base to produce an intermediate result. Finally, the defuzzification module uses this intermediate result to obtain the global output.

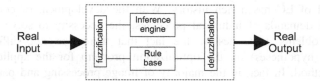

Fig. 1.2 Elements of a fuzzy system

1.3 Neural Networks

Neural networks (NN) [4], are defined as interconnected systems which parallel wise perform a general process. Such elements are undergone to a learning task which automatically modifies the network parameters as a consequence of the optimization of a certain criterion. Each unit of the neural network is considered as a highly simplified model of the biological neuron present in the animal brain.

The main characteristics of NN models can be summarized as follows:

- Parallelism: All neural network architectures maintain some level of parallelism in the computation of their numerous units;
- Units: The basic units of a NN correspond to the same elements which maintain the same characteristics and behavior;
- Information processing: the output value of each unit depend completely on its current state and the output values of other units to which it is associated; and
- Learning: NN parameters are undergone to changes, according to a learning scheme which minimizes a certain performance criteria.

Neural networks architectures are classified in according to several criteria:

1. According to their connections, there are three major types of NN such as, a. recurrent network, b. Feed forward network and c. Competitive networks.
2. According to their structure: (a) static (fixed) structure and (b) dynamic structure.
3. According to their learning algorithms, a. Supervised learning and b. Unsupervised learning.
4. Neural networks models can also be divided regarding the processes in which they are used such as a. Pattern recognition, b. Classification and c. Regression.

1.4 Evolutionary Computation

Evolutionary computation (EC) [5] methods are derivative-free procedures, which do not require that the objective function must be neither two-times differentiable nor uni-modal. Therefore, EC methods as global optimization algorithms can deal with non-convex, nonlinear, and multimodal problems subject to linear or nonlinear constraints with continuous or discrete decision variables.

The field of EC has a rich history. With the development of computational devices and demands of industrial processes, the necessity to solve some optimization problems arose despite the fact that there was not sufficient prior knowledge (hypotheses) on the optimization problem for the application of a classical method. In fact, in the majority of image processing and pattern recognition cases, the problems are highly nonlinear, or characterized by a noisy fitness, or without an explicit analytical expression as the objective function might be the result of an experimental or simulation process. In this context, the EC methods have been proposed as optimization alternatives.

A EC technique is a general method for solving optimization problems. It uses an objective function in an abstract and efficient manner, typically without utilizing deeper insights into its mathematical properties. EC methods do not require hypotheses on the optimization problem nor any kind of prior knowledge on the objective function. The treatment of objective functions as "black boxes" [6] is the most prominent and attractive feature of EC methods.

EC methods obtain knowledge about the structure of an optimization problem by utilizing information obtained from the possible solutions (i.e., candidate solutions) evaluated in the past. This knowledge is used to construct new candidate solutions which are likely to have a better quality.

Recently, several EC methods have been proposed with interesting results. Such approaches use as inspiration our scientific understanding of biological, natural or social systems, which at some level of abstraction can be represented as optimization processes [7]. These methods include the social behavior of bird flocking and fish schooling such as the particle swarm optimization (PSO) algorithm [8], the cooperative behavior of bee colonies such as the artificial bee colony (ABC) technique [9], the improvisation process that occurs when a musician searches for a better state of harmony such as the harmony search (HS) [10], the emulation of the bat behavior such as the bat algorithm (BA) method [11], the mating behavior of firefly insects such as the firefly (FF) method [12], the social-spider behavior such as the social spider optimization (SSO) [13], the simulation of the animal behavior in a group such as the collective animal behavior [14], the emulation of immunological systems as the clonal selection algorithm (CSA) [15], the simulation of the electromagnetism phenomenon as the electromagnetism-Like algorithm [16], and the emulation of the differential and conventional evolution in species such as the Differential Evolution (DE) [17] and genetic algorithms (GA) [18], respectively.

1.5   Definition of an Optimization Problem

The vast majority of image processing and pattern recognition algorithms use some form of optimization, as they intend to find some solution which is "best" according to some criterion. From a general perspective, an optimization problem is a situation that requires to decide for a choice from a set of possible alternatives in order to reach a predefined/required benefit at minimal costs [19].

Consider a public transportation system of a city, for example. Here the system has to find the "best" route to a destination location. In order to rate alternative solutions and eventually find out which solution is "best," a suitable criterion has to be applied. A reasonable criterion could be the distance of the routes. We then would expect the optimization algorithm to select the route of shortest distance as a solution. Observe, however, that other criteria are possible, which might lead to different "optimal" solutions, e.g., number of transfers, ticket price or the time it takes to travel the route leading to the fastest route as a solution.

Mathematically speaking, optimization can be described as follows: Given a function $f : S \rightarrow \mathbb{R}$ which is called the objective function, find the argument which minimizes f:

$$x^* = \arg\min_{x \in S} f(x) \tag{1.1}$$

S defines the so-called solution set, which is the set of all possible solutions for the optimization problem. Sometimes, the unknown(s) x are referred to design variables. The function f describes the optimization criterion, i.e., enables us to calculate a quantity which indicates the "quality" of a particular x.

In our example, S is composed by the subway trajectories and bus lines, etc., stored in the database of the system, x is the route the system has to find, and the optimization criterion $f(x)$ (which measures the quality of a possible solution) could calculate the ticket price or distance to the destination (or a combination of both), depending on our preferences.

Sometimes there also exist one or more additional constraints which the solution x^* has to satisfy. In that case we talk about constrained optimization (opposed to unconstrained optimization if no such constraint exists). As a summary, an optimization problem has the following components:

- One or more design variables x for which a solution has to be found
- An objective function $f(x)$ describing the optimization criterion
- A solution set S specifying the set of possible solutions x
- (Optional) One or more constraints on x

In order to be of practical use, an optimization algorithm has to find a solution in a reasonable amount of time with reasonable accuracy. Apart from the performance of the algorithm employed, this also depends on the problem at hand itself. If we can hope for a numerical solution, we say that the problem is well-posed. For assessing whether an optimization problem is well-posed, the following conditions must be fulfilled:

1. A solution exists.
2. There is only one solution to the problem, i.e., the solution is unique.
3. The relationship between the solution and the initial conditions is such that small perturbations of the initial conditions result in only small variations of x^*.

1.6   Classical Optimization

Once a task has been transformed into an objective function minimization problem, the next step is to choose an appropriate optimizer. Optimization algorithms can be divided in two groups: derivative-based and derivative-free [20].

In general, $f(x)$ may have a nonlinear form respect to the adjustable parameter x. Due to the complexity of $f(\cdot)$, in classical methods, it is often used an iterative algorithm to explore the input space effectively. In iterative descent methods, the next point x_{k+1} is determined by a step down from the current point x_k in a direction vector $\mathbf{d}$:

$$x_{k+1} = x_k + \alpha \mathbf{d}, \qquad (1.2)$$

where α is a positive step size regulating to what extent to proceed in that direction. When the direction d in Eq. (1.2) is determined on the basis of the gradient ($\mathbf{g}$) of the objective function $f(\cdot)$, such methods are known as gradient-based techniques.

The method of steepest descent is one of the oldest techniques for optimizing a given function. This technique represents the basis for many derivative-based methods. Under such a method, the Eq. (1.3) becomes the well-known gradient formula:

$$x_{k+1} = x_k - \alpha \mathbf{g}(f(x)), \qquad (1.3)$$

However, classical derivative-based optimization can be effective as long the objective function fulfills two requirements:

- The objective function must be two-times differentiable.
- The objective function must be uni-modal, i.e., have a single minimum.

A simple example of a differentiable and uni-modal objective function is

$$f(x_1, x_2) = 10 - e^{-\left(x_1^2 + 3 \cdot x_2^2\right)} \qquad (1.4)$$

Figure 1.3 shows the function defined in Eq. (1.4).

Unfortunately, under such circumstances, classical methods are only applicable for a few types of optimization problems. For combinatorial optimization, there is no definition of differentiation.

Furthermore, there are many reasons why an objective function might not be differentiable. For example, the "floor" operation in Eq. (1.5) quantizes the function in Eq. (1.4), transforming Fig. 1.3 into the stepped shape seen in Fig. 1.4. At each step's edge, the objective function is non-differentiable:

$$f(x_1, x_2) = \text{floor}\left(10 - e^{-\left(x_1^2 + 3 \cdot x_2^2\right)}\right) \qquad (1.5)$$

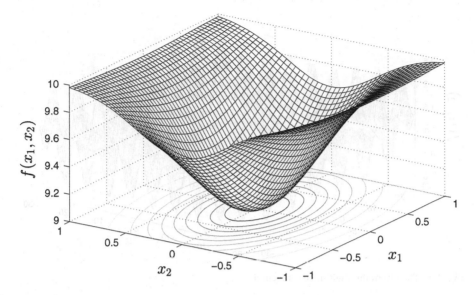

Fig. 1.3 Uni-modal objective function

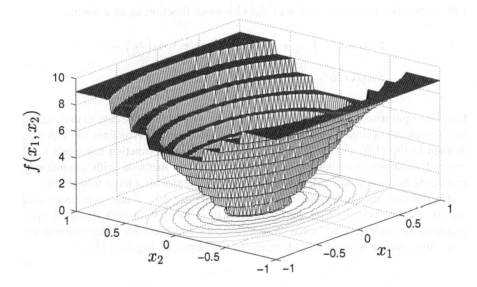

Fig. 1.4 A non-differentiable, quantized, uni-modal function

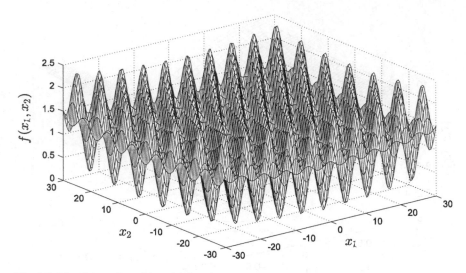

Fig. 1.5 The Griewank multi-modal function

Even in differentiable objective functions, gradient-based methods might not work. Let us consider the minimization of the Griewank function as an example

$$\text{Minimize} \quad f(x_1, x_2) = \frac{x_1^2 + x_2^2}{4000} - \cos(x_1) \cos\left(\frac{x_2}{\sqrt{2}}\right) + 1$$
$$\text{Subject to} \quad \begin{array}{l} -30 \leq x_1 \leq 30 \\ -30 \leq x_2 \leq 30 \end{array} \tag{1.6}$$

From the optimization problem formulated in Eq. (1.6), it is quite easy to understand that the global optimal solution is $x_1 = x_2 = 0$. Figure 1.5 visualizes the function defined in Eq. (1.6). According to Fig. 1.5, the objective function has many local optimal solutions (multimodal) so that the gradient methods with a randomly generated initial solution will converge to one of them with a large probability.

Considering the limitations of gradient-based methods, image processing and pattern recognition problems make difficult their integration with classical optimization methods. Instead, some other techniques which do not make assumptions and which can be applied to wide range of problems are required [3].

1.7   Optimization with Evolutionary Computation

From a conventional point of view, an EC method is an algorithm that simulates at some level of abstraction a biological, natural or social system. To be more specific, a standard EC algorithm includes:

1. One or more populations of candidate solutions are considered.
2. These populations change dynamically due to the production of new solutions.
3. A fitness function reflects the ability of a solution to survive and reproduce.
4. Several operators are employed in order to explore an exploit appropriately the space of solutions.

The EC methodology suggest that, on average, candidate solutions improve their fitness over generations (i.e., their capability of solving the optimization problem). A simulation of the evolution process based on a set of candidate solutions whose fitness is properly correlated to the objective function to optimize will, on average, lead to an improvement of their fitness and thus steer the simulated population towards the global solution.

Most of the optimization methods have been designed to solve the problem of finding a global solution of a nonlinear optimization problem with box constraints in the following form:

$$\text{Maximize} \quad f(\mathbf{x}), \quad \mathbf{x} = (x_1, \dots, x_d) \in \mathbb{R}^d$$
$$\text{Subject to} \quad \mathbf{x} \in \mathbf{X} \tag{1.7}$$

where $f : \mathbb{R}^d \to \mathbb{R}$ is a nonlinear function whereas $\mathbf{X} = \left\{ \mathbf{x} \in \mathbb{R}^d | l_i \le x_i \le u_i, i = 1, \dots, d. \right\}$ is a bounded feasible search space, constrained by the lower (l_i) and upper (u_i) limits.

In order to solve the problem formulated in Eq. (1.6), in an evolutionary computation method, a population $\mathbf{P}^k (\{ \mathbf{p}_1^k, \mathbf{p}_2^k, \dots, \mathbf{p}_N^k \})$ of N candidate solutions (individuals) evolves from the initial point ($k = 0$) to a total *gen* number iterations ($k = gen$). In its initial point, the algorithm begins by initializing the set of N candidate solutions with values that are randomly and uniformly distributed between the pre-specified lower (l_i) and upper (u_i) limits. In each iteration, a set of evolutionary operators are applied over the population $\mathbf{P}^k$ to build the new population $\mathbf{P}^{k+1}$. Each candidate solution $\mathbf{p}_i^k$ ($i \in [1, \dots, N]$) represents a d-dimensional vector $\left\{ p_{i,1}^k, p_{i,2}^k, \dots, p_{i,d}^k \right\}$ where each dimension corresponds to a decision variable of the optimization problem at hand. The quality of each candidate solution $\mathbf{p}_i^k$ is evaluated by using an objective function $f(\mathbf{p}_i^k)$ whose final result represents the fitness value of $\mathbf{p}_i^k$. During the evolution process, the best candidate solution $\mathbf{g}$ ($g_1, g_2, \dots g_d$) seen so-far is preserved considering that it represents the best available solution. Figure 1.6 presents a graphical representation of a basic cycle of a EC method.

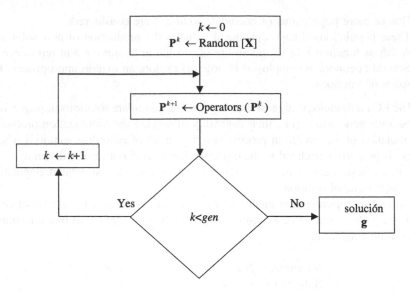

Fig. 1.6 The basic cycle of a EC method

1.8   Soft Computing in Engineering

Recently, there is a considerable interest in the use of soft computing techniques in engineering applications. The high increment of practical examples, in which soft computing has been used, has illustrated its potential as a computational tool.

Currently, we are faced with many complex engineering systems that need to be manipulated. As they are usually not fully theoretically tractable, it is not possible to use traditional deterministic methods. Soft computing, as opposed to conventional deterministic methods, is a set of methodologies (working synergistically, not competitively) that, in one form or another, exploits the tolerance for imprecision, uncertainty and approximate reasoning to achieve tractability, robustness, low-cost solution, and close resemblance with human-like decision making. Soft computing techniques include neural networks, evolutionary computation, fuzzy logic and Learning Automata. The recent years have witnessed tremendous success of these powerful methods in virtually all areas of science and technology, as evidenced by the large numbers of research results published in a variety of journals, conferences, as well as many books.

Engineering is a rich source of problems where each new approach that is developed by mathematicians and computer scientists is quickly identified, understood and assimilated in order to be applied to specific problems. In this book we strive to bring some state of the art techniques by using recent Soft computing techniques after its application to challenging and significant problems in engineering.

Soft computing methods have many variants. There exist a rich amount of literature on the subject, including textbooks, tutorials, and journal papers that cover in detail practically every aspect of the field. The great amount of information available makes it difficult for no specialist to explore the literature and to find the right technique for a specific engineering application. Therefore, any attempt to present the whole area of Soft Computing in detail would be a daunting task, probably doomed to failure. This task would be even more difficult if the goal is to understand the applications of Soft Computing techniques in the context of engineering application. For this reason, the best practice is to consider only a representative set of Soft Computing approaches, just as it has been done in this book.

References

1. Patnaik S, Zhong B (eds) (2014) Soft computing techniques in engineering applications, studies in computational intelligence. Springer Verlag, Berlin
2. Chaturvedi DK (2008) Soft computing techniques and its applications in electrical engineering, studies in computational intelligence, 103. Springer Verlag, New York
3. Ghosh S, Razouqi Q, Schumacher HJ, Celmins A (1998) A survey of recent advances in fuzzy logic in telecommunications networks and new challenges. IEEE Trans Fuzzy Syst 6(3):443–447
4. Zhang GP (2000) Neural networks for classification: a survey. IEEE Trans Syst Man Cybern Part C Appl Rev 30(4):451–462
5. Dan S (2013) Evolutionary optimization algorithms. Wiley, New York
6. Blum C, Roli A (2003) Metaheuristics in combinatorial optimization: overview and conceptual comparison. ACM Comput Surv (CSUR) 35(3):268–308. doi:10.1145/937503. 937505
7. Satyasai JN, Panda G (2014) A survey on nature inspired metaheuristic algorithms for partitional clustering. Swarm Evol Comput 16:1–18
8. Kennedy J, Eberhart R (1995) Particle swarm optimization. Proceed 1995 IEEE Int Confere Neural Networks 4:1942–1948 (December 1995)
9. Karaboga D (2005) An idea based on honey bee swarm for numerical optimization. TechnicalReport-TR06. Engineering Faculty, Computer Engineering Department, Erciyes University
10. Geem ZW, Kim JH, Loganathan GV (2001) A new heuristic optimization algorithm: harmony search. Simulations 76:60–68
11. Yang XS (2010) A new metaheuristic bat-inspired algorithm. In: Cruz C, González J, Krasnogor GTN, Pelta DA (eds) Nature inspired cooperative strategies for optimization (NISCO 2010), studies in computational intelligence, vol 284. Springer Verlag, Berlin, pp 65–74
12. Yang XS (2009) Firefly algorithms for multimodal optimization. In: Stochastic algorithms: foundations and applications, SAGA 2009, Lecture Notes in Computer Sciences, vol. 5792, pp 169–178
13. Cuevas Erik, Cienfuegos Miguel, Zaldívar Daniel, Pérez-Cisneros Marco (2013) A swarm optimization algorithm inspired in the behavior of the social-spider. Expert Syst Appl 40 (16):6374–6384
14. Cuevas E, González M, Zaldivar D, Pérez-Cisneros M, García G (2012) An algorithm for global optimization inspired by collective animal behaviour. Dis Dyn Nat Soc 638275

15. de Castro LN, von Zuben FJ (2002) Learning and optimization using the clonal selection principle. IEEE Trans Evol Comput 6(3):239–251
16. Birbil ŞI, Fang SC (2003) An electromagnetism-like mechanism for global optimization. J Glob Optim 25(1):263–282
17. Storn R, Price K (1995) Differential evolution—a simple and efficient adaptive scheme for global optimisation over continuous spaces. TechnicalReportTR-95–012, ICSI, Berkeley, CA
18. Goldberg DE (1989) Genetic algorithm in search optimization and machine learning. Addison-Wesley, Boston
19. Akay Bahriye, Karaboga Dervis (2015) A survey on the applications of artificial bee colony in signal, image, and video processing. SIViP 9(4):967–990
20. Xin-She Y (2010) Engineering optimization. Wiley, USA

Chapter 2
Motion Estimation Algorithm Using Block-Matching and Harmony Search Optimization

Motion estimation is one of the major problems in developing video coding applications. Motion estimation is one of the major problems in developing video coding applications. On the other hand, block-matching (BM) algorithms are the most popular methods due to their effectiveness and simplicity for both software and hardware implementations. A BM approach assumes that the movement of pixels within a defined region of the current frame can be modeled as a translation of pixels contained in the previous frame. During this procedure is obtained a motion vector by minimizing a certain matching metric that is produced between the current frame and the previous frame. However, the evaluation of such matching measurement is computationally expensive and represents the most consuming operation in the BM process. Therefore, BM motion estimation can be viewed as an optimization problem whose goal is to find the best-matching block within a search space. Harmony search (HS) algorithm is a metaheuristic optimization method inspired by the music improvisation process, in which a musician polishes the pitches to obtain a better state of harmony. In this chapter, a BM algorithm that combines HS with a fitness approximation model is presented. The approach uses motion vectors belonging to the search window as potential solutions. A fitness function evaluates the matching quality of each motion vector candidate. In order to minimize computational time, the approach incorporates a fitness calculation strategy to decide which motion vectors can be only estimated or actually evaluated. Guided by the values of such a fitness calculation strategy, the set of motion vectors is evolved through HS operators until the best possible motion vector is identified. The presented method has been compared to other BM algorithms in terms of velocity and coding quality and its experimental results demonstrate that the algorithm exhibits the best balance between coding efficiency and computational complexity.

© Springer International Publishing AG 2017
M.-A. Díaz-Cortés et al., *Engineering Applications of Soft Computing*,
Intelligent Systems Reference Library 129, DOI 10.1007/978-3-319-57813-2_2

2.1   Introduction

Motion estimation plays important roles in a number of applications such as automobile navigation, video coding, surveillance cameras and so forth. The measurement of the motion vector is a fundamental problem in image processing and computer vision, which has been faced using several approaches [1–4]. The goal is to compute an approximation to the 2-D motion field—a projection of the 3-D velocities of surface points onto the imaging surface.

Video coding is currently utilized in a vast amount of applications ranging from fixed and mobile telephony, real-time video conferencing, DVD and high-definition digital television. Motion estimation (ME) is an important part of any video coding system, since it can achieve significant compression by exploiting the temporal redundancy existing in a video sequence. Several ME methods have been studied aiming for a complexity reduction at video coding, such as block matching (BM) algorithms, parametric-based models [5], optical flow [6] and pel-recursive techniques [7]. Among such methods, BM seems to be the most popular technique due to its effectiveness and simplicity for both software and hardware implementations [8]. Furthermore, in order to reduce the computational complexity in ME, many BM algorithms have been proposed and used in implementations of various video compression standards such as MPEG-4 [9] and H.264 [10].

In BM algorithms, the video frames are partitioned in non-overlapping blocks of pixels. Each block is predicted from a block of equal size in the previous frame. Specifically, for each block in the current frame, we search for a best matching block within a searching window in the previous frame that minimizes a certain matching metric. The most used matching measure is the sum of absolute differences (SAD) which is computationally expensive and represents the most consuming operation in the BM process. The best matching block found represents the predicted block, whose displacement from the previous block is represented by a transitional motion vector (MV). Therefore, BM is essentially an optimization problem, with the goal of finding the best matching block within a search space.

The full search algorithm (FSA) [11] is the simplest block-matching algorithm that can deliver the optimal estimation solution regarding a minimal matching error as it checks all candidates one at a time. However, such exhaustive search and full-matching error calculation at each checking point yields an extremely computational expensive BM method that seriously constraints real-time video applications.

In order to decrease the computational complexity of the BM process, several BM algorithms have been proposed considering the following three techniques: (1) using a fixed pattern: which means that the search operation is conducted over a fixed subset of the total search window. The three step search (TSS) [12], the new three step search (NTSS) [13], the simple and efficient TSS (SES) [14], the four step search (4SS) [15] and the diamond search (DS) [16] are some of its well-known examples. Although such approaches have been algorithmically considered as the fastest, they are not able eventually to match the dynamic motion-content, delivering false motion vectors (image distortions). (2) Reducing the search points: in this method, the algorithm chooses as search points exclusively those locations which iteratively minimize the error-function (SAD values). This category includes: the adaptive rood pattern search (ARPS) [17], the fast block matching using prediction (FBMAUPR) [18], the block-based gradient descent search (BBGD) [19] and the neighbourhood elimination algorithm (NE) [20]. Such approaches assume that the error-function behaves monotonically, holding well for slow-moving sequences; however, such properties do not hold true for other kind of movements in video sequences [21], which risks on algorithms getting trapped into local minima. (3) Decreasing the computational overhead for every search point, which means the matching cost (SAD operation) is replaced by a partial or a simplify version that features less complexity. The new pixel-decimation (ND) [22], the efficient block matching using multilevel intra and inter-sub-blocks [13] and the successive elimination algorithm [23], all assume that all pixels within each block move by the same amount and a good estimate of the motion could be obtained through only a fraction of the pixel pool. However, since only a fraction of pixels enters into the matching computation, the use of these regular sub-sampling techniques can seriously affect the accuracy of the detection of motion vectors due to noise or illumination changes.

Another popular group of BM algorithms employ spatiotemporal correlation, using the neighboring blocks in spatial and temporal domain. The main advantage of these algorithms is that they alleviate the local minimum problem to some extent. Since the new initial or predicted search center is usually closer to the global minimum, the chance of getting trapped in a local minimum decreases. This idea has been incorporated by many fast block motion estimation algorithms such as the enhanced predictive zonal search (EPZS) [24] and the UMHexagonS [25]. However, the information delivered by the neighboring blocks occasionally conduces to false initial search points, producing distorted motion vectors. Such problem is typically caused when very small objects moves during the image sequence [26].

Alternatively, evolutionary approaches such as genetic algorithms (GA) [27] and particle swarm optimization (PSO) [28] are well known for locating the global optimum in complex optimization problems. Despite of such fact, only few evolutionary approaches have specifically addressed the problem of BM, such as the light-weight genetic block matching (LWG) [29], the genetic four-step search (GFSS) [30] and the PSO-BM [31]. Although these methods support an accurate identification of the motion vector, their spending times are very long in comparison to other BM techniques.

On the other hand, the harmony search (HS) algorithm introduced by Geem et al. [32] is one of the population-based evolutionary heuristics algorithms which are based on the metaphor of the improvisation process that occurs when a musician searches for a better state of harmony. The HS generates a new candidate solution from all existing solutions. In HS, the solution vector is analogous to the harmony in music, and the local and global search schemes are analogous to musician's improvisations. In comparison to other meta-heuristics in the literature, HS imposes fewer mathematical requirements as it can be easily adapted for solving several sorts of engineering optimization challenges [33, 34]. Furthermore, numerical comparisons have demonstrated that the evolution for the HS is faster than GA [33, 35, 36], attracting ever more attention. It has been successfully applied to solve a wide range of practical optimization problems such as structural optimization, parameter estimation of the nonlinear Muskingum model, design optimization of water distribution networks, vehicle routing, combined heat and power economic dispatch, design of steel frames, bandwidth-delay-constrained least-cost multicast routing, computer vision, among others [35–43].

A main difficulty applying HS to solve real-world problems is that it usually needs a large number of fitness evaluations before an acceptable result can be obtained. In practice, however, fitness evaluations are not always straightforward because either an explicit fitness function does not exist (an experiment is needed instead) or the evaluation of the fitness function is computationally demanding. Furthermore, since random numbers are involved in the calculation of new individuals, they may encounter the same positions (repetition) that other individuals have visited in previous iterations, especially when the individuals are confined to a small area.

The problem of considering expensive fitness evaluations has already been faced in the field of evolutionary algorithms (EA) and is better known as fitness approximation [44]. In such approach, the idea is to estimate the fitness value of so many individuals as it is possible instead of evaluating the complete set. Such estimations are based on an approximate model of the fitness landscape. Thus, the individuals to be evaluated and those to be estimated are determined following some fixed criteria which depend on the specific properties of the approximate model [45]. The models involved at the estimation can be built during the actual EA run, since EA repeatedly sample the search space at different points [46]. There are many possible approximation models and several have already been used in combination with EA (e.g. polynomials [47], the kriging model [48], the feed-forward neural networks that includes multi-layer perceptrons [49] and radial basis-function networks [50]). These models can be either global, which make use of all available data or local which make use of only a small set of data around the point where the function is to be approximated. Local models, however, have a number of advantages [46]: they are well-known and suitably established techniques with relatively fast speeds. Moreover, they consider the intuitively most important information: the closest neighbors.

In this chapter, is presented a BM algorithm that combines HS with a fitness approximation model. Since the presented method approaches the BM process as an optimization problem, its overall operation can be formulated as follows: First, a population of individuals is initialized where each individual represents a motion vector candidate (a search location). Then, the set of HS operators is applied at each iteration in order to generate a new population. The procedure is repeated until convergence is reached whereas the best solution is expected to represent the most accurate motion vector. In the optimization process, the quality of each individual is evaluated through a fitness function which represents the SAD value corresponding to each motion vector. In order to save computational time, the approach incorporates a fitness estimation strategy to decide which search locations can be only estimated or actually evaluated. The method has been compared to other BM algorithms in terms of velocity and coding quality. Experimental results show that the HS-BM algorithm exhibits the best trade-off between coding efficiency and computational complexity.

The overall chapter is organized as follows: Sect. 2.2 holds a brief description about the HS method in Sect. 2.3, the fitness calculation strategy for solving the expensive optimization problem is presented. Section 2.4 provides background about the BM motion estimation issue while Sect. 2.5 exposes the final BM algorithm as a combination of HS and the fitness calculation strategy. Section 2.6 demonstrates experimental results for the presented approach over standard test sequences and some conclusions are drawn in Sect. 2.7.

2.2 Harmony Search Algorithm

2.2.1 The Harmony Search Algorithm

In the basic HS, each solution is called a "harmony" and is represented by an n-dimension real vector. An initial population of harmony vectors are randomly generated and stored within a harmony memory (HM). A new candidate harmony is thus generated from the elements in the HM by using a memory consideration operation either by a random re-initialization or a pitch adjustment operation. Finally, the HM is updated by comparing the new candidate harmony and the worst harmony vector in the HM. The worst harmony vector is replaced by the new candidate vector in case it is better than the worst harmony vector in the HM. The above process is repeated until a certain termination criterion is met. The basic HS algorithm consists of three basic phases: HM initialization, improvisation of new harmony vectors and updating of the HM. The following discussion addresses details about each stage.

2.2.1.1 Initializing the Problem and Algorithm Parameters

In general, the global optimization problem can be summarized as follows: min $f(\mathbf{x})$: $x(j) \in [l(j), u(j)], j = 1, 2, \ldots, n$, where $f(\mathbf{x})$ is the objective function, $\mathbf{x} = (x(1), x(2), \ldots, x(n))$ is the set of design variables, n is the number of design variables, and $l(j)$ and $u(j)$ are the lower and upper bounds for the design variable $x(j)$, respectively. The parameters for HS are the harmony memory size, i.e., the number of solution vectors lying on the harmony memory (HM), the harmony-memory consideration rate ($HMCR$), the pitch adjusting rate (PAR), the distance bandwidth (BW) and the number of improvisations (NI) which represents the total number of iterations. It is obvious that the performance of HS is strongly influenced by parameter values which determine its behavior.

2.2.1.2   Harmony Memory Initialization

In this stage, initial vector components at HM, i.e., HMS vectors, are configured. Let $\mathbf{x}_i = \{x_i(1), x_i(2), \ldots, x_i(n)\}$ represent the i-th randomly-generated harmony vector: $x_i(j) = l(j) + (u(j) - l(j)) \cdot \text{rand}(0,1)$ for $j = 1, 2, \ldots, n$ and $i = 1, 2, \ldots,$ HMS, where $\text{rand}(0,1)$ is a uniform random number between 0 and 1. Then, the HM matrix is filled with the HMS harmony vectors as follows:

$$\text{HM} = \begin{bmatrix} \mathbf{x}_1 \\ \mathbf{x}_2 \\ \vdots \\ \mathbf{x}_{HMS} \end{bmatrix} \tag{2.1}$$

2.2.1.3   Improvisation of New Harmony Vectors

In this phase, a new harmony vector $\mathbf{x}_{new}$ is built by applying the following three operators: memory consideration, random re-initialization and pitch adjustment. Generating a new harmony is known as 'improvisation'. In the memory consideration step, the value of the first decision variable $x_{new}(1)$ for the new vector is chosen randomly from any of the values already existing in the current HM i.e., from the set $\{x_1(1), x_2(1), \ldots, x_{HMS}(1)\}$. For this operation, a uniform random number r_1 is generated within the range [0, 1]. If r_1 is less than $HMCR$, the decision variable $x_{new}(1)$ is generated through memory considerations; otherwise, $x_{new}(1)$ is obtained from a random re-initialization between the search bounds $[l(1), u(1)]$. Values of the other decision variables $x_{new}(2), x_{new}(3), \ldots, x_{new}(n)$ are also chosen accordingly. Therefore, both operations, memory consideration and random re-initialization, can be modelled as follows:

$$x_{new}(j) = \begin{cases} x_i(j) \in \{x_1(j), x_2(j), \ldots, x_{HMS}(j)\} & \text{with probability } HMCR \\ l(j) + (u(j) - l(j)) \cdot \text{rand}(0,1) & \text{with probability } 1\text{-}HMCR \end{cases} \tag{2.2}$$

Every component obtained by memory consideration is further examined to determine whether it should be pitch-adjusted. For this operation, the pitch-adjusting rate (PAR) is defined as to assign the frequency of the adjustment and the bandwidth factor (BW) to control the local search around the selected elements of the HM. Hence, the pitch adjusting decision is calculated as follows:

$$x_{new}(j) = \begin{cases} x_{new}(j) = x_{new}(j) \pm \text{rand}(0,1) \cdot BW & \text{with probability } PAR \\ x_{new}(j) & \text{with probability } (1\text{-}PAR) \end{cases} \tag{2.3}$$

Pitch adjusting is responsible for generating new potential harmonies by slightly modifying original variable positions. Such operation can be considered similar to the mutation process in evolutionary algorithms. Therefore, the decision variable is either perturbed by a random number between 0 and BW or left unaltered. In order to protect the pitch adjusting operation, it is important to assure that points lying outside the feasible range $[l, u]$ must be re-assigned i.e., truncated to the maximum or minimum value of the interval.

2.2.1.4 Updating the Harmony Memory

After a new harmony vector x_{new} is generated, the harmony memory is updated by the survival of the fit competition between x_{new} and the worst harmony vector x_w in the HM. Therefore x_{new} will replace x_w and become a new member of the HM in case the fitness value of x_{new} is better than the fitness value of x_w.

2.2.2 Computational Procedure

The computational procedure of the basic HS can be summarized as follows [18]:

Step 1:	Set the parameters *HMS*, *HMCR*, *PAR*, *BW* and *NI*
Step 2:	Initialize the HM and calculate the objective function value of each harmony vector

Step 3:	Improvise a new harmony $\mathbf{x}_{new}$ as follows:
	for ($j = 1$ to n) do
	    if ($r_1 < HMCR$) then
	        $x_{new}(j) = x_a(j)$ where a is element of $(1, 2, \ldots, HMS)$ randomly selected
	        if ($r_2 < PAR$) then
	            $x_{new}(j) = x_{new}(j) \pm r_3 \cdot BW$ where $r_1, r_2, r_3 \in \text{rand}(0, 1)$
	        end if
	        if $x_{new}(j) < l(j)$
	            $x_{new}(j) = l(j)$
	        end if
	        if $x_{new}(j) > u(j)$
	            $x_{new}(j) = u(j)$
	        end if
	    else
	        $x_{new}(j) = l(j) + r \cdot (u(j) - l(j))$, where $r \in \text{rand}(0, 1)$
	    end if
	end for

Step 4:	Update the *HM* as $\mathbf{x}_w = \mathbf{x}_{new}$ if $f(\mathbf{x}_{new}) < f(\mathbf{x}_w)$
Step 5:	If *NI* is completed, the best harmony vector $\mathbf{x}_b$ in the HM is returned; otherwise go back to step 3

2.3 Fitness Approximation Method

Evolutionary algorithms that use fitness approximation aim to find the global minimum of a given function considering only a very few number of function evaluations and a large number of estimations, based on an approximate model of the function landscape. In order to apply such approach, it is necessary that the objective function implicates a very expensive evaluation and consists of few dimensions (up to five) [51]. Recently, several fitness estimators have been reported in the literature [47–50] in which the number of function evaluations is considerably reduced to hundreds, dozens, or even less. However, most of these methods produce complex algorithms whose performance is conditioned to the quality of the training phase and the learning algorithm in the construction of the approximation model.

In this chapter, we explore the use of a local approximation scheme based on the nearest-neighbor-interpolation (NNI) for reducing the function evaluation number. The model estimates the fitness values based on previously evaluated neighboring individuals which have been stored during the evolution process. At each generation, some individuals of the population are evaluated through the accurate (real) fitness function while the other remaining individuals are only estimated. The positions to be accurately evaluated are determined based on their proximity to the best individual or regarding their uncertain fitness value.

2.3.1 Updating the Individual Database

In a fitness approximation method, every evaluation of an individual produces one data point (individual position and fitness value) that is potentially taken into account for building the approximation model during the evolution process. Therefore, in our approach, we keep all seen-so-far evaluated individuals and their respective fitness values within a history array **T** which is employed to select the closest neighbor and to estimate the fitness value of a new individual. Thus, each element of **T** consists of two parts: the individual position and its respective fitness value. The array **T** begins with null elements in the first iteration. Then, as the optimization process evolves, new elements are added. Since the goal of a fitness approximation approach is to evaluate the least possible number of individuals, only few elements are contained in **T**.

2.3.2 Fitness Calculation Strategy

This section explains the strategy to decide which individuals are to be evaluated or estimated. The presented fitness calculation scheme estimates most of fitness values to reduce the computational overhead at each generation. In the model, those individuals positioned nearby the individual with the best fitness value at the array **T** (Rule 1) are evaluated by using the actual fitness function. Such individuals are important as they possess a stronger influence over the evolution process than the others. Moreover, it also evaluates those individuals placed in regions of the search space with no previous evaluations (Rule 2). Fitness values for these individuals are uncertain since there is no close reference (close points contained in **T**) to calculate their estimates.

The remaining individuals, for which there exist a close point that is previously evaluated and its fitness value is not the best contained in the array **T**, are estimated using the NNI criterion (Rule 3). Thus, the fitness value of an individual is approximated by assigning the same fitness value that the nearest individual stored in **T**.

Therefore, the fitness computation model follows three important rules to evaluate or estimate fitness values:

1. *Exploitation rule (evaluation).* If a new individual (search position) P is located closer than a distance d with respect to the nearest individual L_q contained in **T** ($q = 1, 2, 3, \ldots, m$; where m is the number of elements contained in **T**), whose fitness value F_{L_q} corresponds to the best fitness value, then the fitness value of P is evaluated by using the actual fitness function. Figure 2.1a draws the rule procedure.

2. *Exploration rule (evaluation).* If a new individual P is located further away than a distance d with respect to the nearest individual L_q contained in **T**, then its fitness value is evaluated by using the actual fitness function. Figure 2.1b outlines the rule procedure.

3. *NNI rule (estimation).* If a new individual P is located closer than a distance d with respect to the nearest individual L_q contained in **T**, whose fitness value F_{L_q} does not correspond to the best fitness value, then its fitness value is estimated assigning it the same fitness that $L_q (F_P = F_{L_q})$. Figure 2.1c sketches the rule procedure.

The d value controls the trade-off between the evaluation and the estimation of search locations. Typical values of d range from 1 to 4. Values close to 1 improve the precision at the expense of a higher number of fitness evaluations (the number of evaluated individuals is more than the number of estimated). On the other hand, values close to 4 decrease the computational complexity at the price of poor accuracy (decreasing the number of evaluation and increasing the number of estimations). After exhaustive experimentation, it has been determined that a value of $d = 3$ represents the best trade-off between computational overhead and accuracy, so it is used throughout the study. The presented method, from an optimization perspective, favors the exploitation and exploration in the search process. For the exploration, the method evaluates the fitness function of new search locations which have been located far from previously calculated positions. Additionally, it also estimates those that are closer. For the exploitation, the presented method

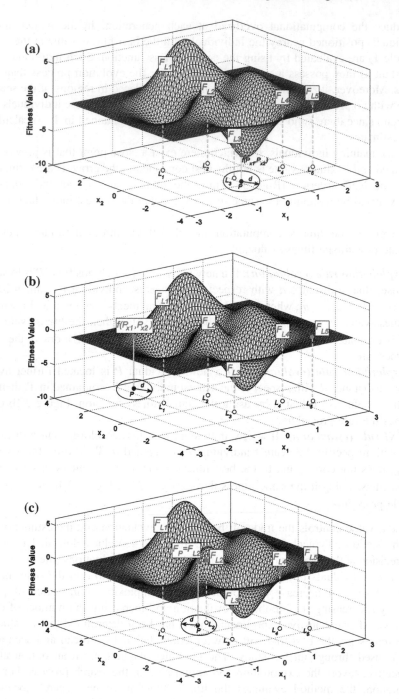

◀**Fig. 2.1** The fitness calculation strategy. **a** According to the rule 1, the individual (search position) P is evaluated since it is located closer than a distance d with respect to the nearest individual location L_3. Therefore, the fitness value F_{L_3} corresponds to the best fitness value (minimum). **b** According to the rule 2, the search point P is evaluated and there is no close reference within its neighborhood. **c** According to rule 3, the fitness value of P is estimated by means of the NNI-estimator, assigning $F_P = F_{L_2}$

evaluates the fitness function of those new searching locations which are placed nearby the position that holds the minimum fitness value seen so far. Such fact is considered as an strong evidence that the new location could improve the "best value" (the minimum) already found.

With the purpose of knowing which rule must be applied by the fitness approximation strategy, considering a new search position P, it is only necessary to identify the closest individual L_q which is contained in T. Then, it is inquired if the positional relationship between them and the fitness value F_{L_q} of L_q fulfill the properties imposed by each rule (distance, if F_{L_q} is the best fitness values contained in T, etc.). As the number of elements of the array **T** is very limited, the computational complexity resulting from such operations is negligible. Figure 2.1 illustrates the procedure of fitness computation for a new solution (point P). In the problem, the objective function f is minimized with respect to two parameters (x_1, x_2). In all figures (Fig. 2.1a–c), the individual database array **T** contains five different elements $(L_1, L_2, L_3, L_4, L_5)$ with their corresponding fitness values $(F_{L_1}, F_{L_2}, F_{L_3}, F_{L_4}, F_{L_5})$. Figure 2.1a, b show the fitness evaluation ($f(x_1, x_2)$) of the new solution P, following the rule 1 and 2 respectively, whereas Fig. 2.1c present the fitness estimation of P using the NNI approach which is laid by rule 3.

2.3.3 HS Optimization Method

The coupling of HS and the fitness approximation strategy is presented in this chapter as an optimization approach. The only difference between the conventional HS and the enhanced HS method is the fitness calculation scheme. In the presented algorithm, only some individuals are actually evaluated (Rules 1 and 2) at each generation. All other fitness values for the rest are estimated using the NNI-approach (Rule 3). The estimation is executed by using the individuals previously calculated which are contained in the array **T**.

Figure 2.2 shows the difference between the conventional HS and the presented version. It is clear that two new blocks have been added, the fitness estimation and the updating individual database. Both elements and the actual evolution block, represent the fitness calculation strategy just as it has been explained at Sect. 3.2. As a result, the HS approach can substantially reduce the number of function evaluations preserving the good search capabilities of HS.

Fig. 2.2 Differences between
the conventional HS and the
HS optimization method
presented in this chapter.
a Conventional HS and **b** the
HS algorithm including the
fitness calculation strategy

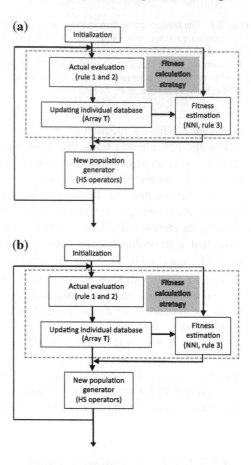

2.4   Motion Estimation and Block-Matching

For motion estimation, in a BM algorithm, the current frame of an image sequence I_t is divided into non-overlapping blocks of $N \times N$ pixels. For each template block in the current frame, the best matched block within a search window (S) of size $(2W + 1) \times (2W + 1)$ in the previous frame I_{t-1} is determined, where W is the maximum allowed displacement. The position difference between a template block in the current frame and the best matched block in the previous frame is called the motion vector (MV) (see Fig. 2.3). Therefore, BM can be viewed as an optimization problem, with the goal of finding the best MV within a search space.

The most well-known matching criterion for BM algorithms is the sum of absolute difference (SAD). It is defined in Eq. (2.5) considering a template block at

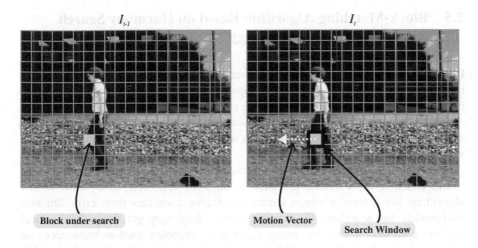

Fig. 2.3 Block matching procedure

position (x, y) in the current frame and the candidate block at position $(x + \hat{u}, y + \hat{v})$ in the previous frame I_{t-1}

$$\text{SAD}(\hat{u}, \hat{v}) = \sum_{j=0}^{N-1} \sum_{i=0}^{N-1} |g_t(x+i, y+j) - g_{t-1}(x+\hat{u}+i, y+\hat{v}+j)| \qquad (2.4)$$

where $g_t(\cdot)$ is the gray value of a pixel in the current frame I_t and $g_{t-1}(\cdot)$ is the gray level of a pixel in the previous frame I_{t-1}. Therefore, the MV in (u, v) is defined as follows:

$$(u, v) = \arg \min_{(u,v) \in S} \text{SAD}(\hat{u}, \hat{v}) \qquad (2.5)$$

where $S = \{(\hat{u}, \hat{v}) | -W \leq \hat{u}, \hat{v} \leq W \text{ and } (x+\hat{u}, y+\hat{v}) \text{ is a valid pixel position } I_{t-1}\}$. As it can be seen, the computing of such matching criterion is a consuming time operation which represents the bottle-neck in the BM process.

In the context of BM algorithms, the FSA is the most robust and accurate method to find the MV. It tests all possible candidate blocks from I_{t-1} within the search area to find the block with minimum SAD. For the maximum displacement of W, the FSA requires $(2W + 1)^2$ search points. For instance, if the maximum displacement W is ± 7, the total search points are 225. Each SAD calculation requires $2N^2$ additions and the total number of additions for the FSA to match a 16×16 block is 130,560. Such computational requirement makes the application of FSA difficult for real time applications.

2.5   Block-Matching Algorithm Based on Harmony Search with the Estimation Strategy

FSA finds the global minimum (the accurate MV), considering all locations within the search space S. Nevertheless, the approach has a high computational cost for practical use. In order to overcome such a problem, many fast algorithms have been developed yielding only a poorer precision than the FSA. A better BM algorithm should spend less computational time on searching and obtaining accurate motion vectors (MVs).

The BM algorithm presented at this chapter is comparable to the fastest algorithms and delivers a similar precision to the FSA approach. Since most of fast algorithms use a regular search pattern or assume a characteristic error function (uni-modal) for searching the motion vector, they may get trapped into local minima considering that for many cases (i.e., complex motion sequences) an uni-modal error is no longer valid. Figure 2.4 shows a typical error surface (SAD values) which has been computed around the search window for a fast-moving sequence. On the other hand, the presented BM algorithm uses a non-uniform search pattern for locating global minimum distortion. Under the effect of the HS operators, the search locations vary from generation to generation, avoiding to get trapped into a local minimum. Besides, since the presented algorithm uses a fitness calculation strategy for reducing the evaluation of the SAD values, it requires fewer search positions.

In the algorithm, the search space S consists of a set of 2-D motion vectors $\hat{u}$ and $\hat{v}$ representing the x and y components of the motion vector, respectively. The particle is defined as:

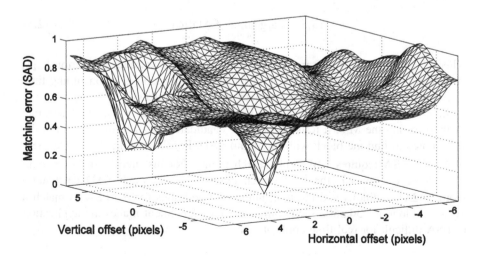

Fig. 2.4 Common non-uni-modal error surface with multiple local minimum error points

$$P_i = \{\hat{u}_1, \hat{v}_i | -W \le \hat{u}_1, \hat{v}_i \le W\} \tag{2.6}$$

where each particle i represents a possible motion vector. In this chapter, the search windows, considered in the simulations, are set to ± 8 and ± 16 pixels. Both configurations are selected in order to compare the results with other approaches presented in the literature.

2.5.1   Initial Population

The first step in HS optimization is to generate an initial group of individuals. The standard literature of evolutionary algorithms generally suggests the use of random solutions as the initial population, considering the absence of knowledge about the problem [52]. However, several studies [53–56] have demonstrated that the use of solutions generated through some domain knowledge to set the initial population (i.e., non-random solutions) can significantly improve its performance. In order to obtain appropriate initial solutions (based on knowledge), an analysis over the motion vector distribution was conducted. After considering several sequences (see Table 2.1; Fig. 2.9), it can be seen that 98% of the MVs are found to lie at the origin of the search window for a slow-moving sequence such as the one at *Container*, whereas complex motion sequences, such as the *Carphone* and the *Foreman* examples, have only 53.5 and 46.7% of their MVs in the central search region. The *Stefan* sequence, showing the most complex motion content, has only 36.9%. Figure 2.5 shows the surface of the MV distribution for the *Foreman* and the *Stefan*. On the other hand, although it is less evident, the MV distribution of several sequences shows small peaks at some locations lying away from the center as they are contained inside a rectangle that is shown in Fig. 2.5b, d by a white overlay. Real-world moving sequences concentrate most of the MVs under a limit due to the motion continuity principle [16]. Therefore, in this chapter, initial solutions are selected from five fixed locations which represent points showing the higher concentration in the MV distribution, just as it is shown by Fig. 2.6.

Table 2.1 Test sequences used in the comparison test

Sequence	Format	Total frames	Motion type
Container	QCIF (176 × 144)	299	Low
Carphone	QCIF (176 × 144)	381	Medium
Foreman	QCIF (352 × 288)	398	Medium
Akiyo	QCIF (352 × 288)	211	Medium
Stefan	CIF (352 × 288)	89	High
Football	SIF (352 × 240)	300	High

(a)

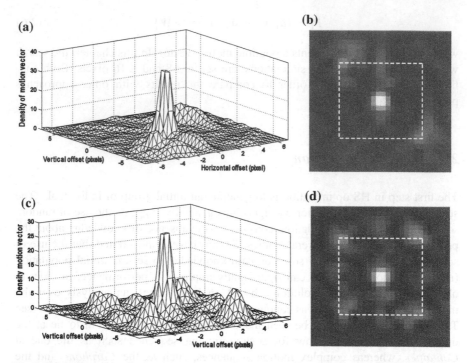

(c)

(b)

(d)

Fig. 2.5 Motion vector distribution for *Foreman* and Stefan sequences. **a, b** MV distribution for the *Foreman* sequence. **c, d** MV distribution for the *Stefan* sequence

Fig. 2.6 Fixed pattern of five elements in the search window of ±8, used as initial solutions

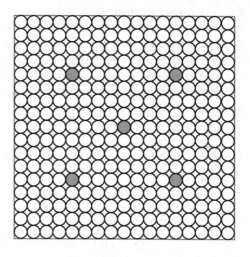

Since most movements suggest displacements near to the center of the search window [12, 13], the initial solutions shown by Fig. 2.6 are used as initial position for the HS algorithm. This consideration is taken regardless of the search window employed (± 8 or ± 16).

2.5.2 Tuning of the HS Algorithm

The performance of HS is strongly influenced by parameter values which determine its behavior. HS incorporates several parameters such as the population size, the operator's probabilities (as *HMCR* and *PAR*) or the total number of iterations (*NI*). Determining the most appropriate parameter values for a determined problem is a complex issue, since such parameters interact to each other in a highly nonlinear manner and there are not mathematical models of such interaction. Throughout the years, two main types of methods have been proposed for setting up parameter values of an evolutionary algorithm: off-line and on-line strategies [57]. An off-line method (called tuning) searches for the best set of parameter values through experimentation. Once defined, these values remain fixed. Such methodology is appropriate when the optimization problem maintains the same properties (dimensionality, multimodality, unconstrained, etc.) each time that the EA is applied. On the other hand, on-line methods focus on changing parameter values during the search process of the algorithm. Thus, the strategy must decide when to change parameter values and determine new values. Therefore, these methods are indicated when EA faces optimizations problems with dimensional variations or restriction changes, etc.

Considering that the optimization problem outlined by the BM process maintains the same properties (same dimensions and similar error landscapes), the off-line method has been used for tuning the HS algorithm. Therefore, after exhaustive experimentation, the following parameters have been found as the best parameter set, $HMCR = 0.7$, $PAR = 0.3$. Considering that the presented approach is tested by using two different search windows (± 8 and ± 16), the values of *BW* and *NI* have different configurations depending on the selected search window. Therefore, it is employed $BW = 8$ and $NI = 25$ in the case of a window search of ± 8 whereas the case of ± 16, it uses $BW = 16$ and $NI = 45$. Once such configurations are defined, the parameter set is kept for all test sequences through all experiments.

2.5.3 The HS-BM Algorithm

The goal of our BM-approach is to reduce the number of evaluations of the SAD values (real fitness function) avoiding any performance loss and achieving an acceptable solution. The HS-BM method is listed below:

Step 1:	Set the HS parameters. $HMCR = 0.7$, $PAR = 0.3$, $BW = 8$ in case of a search window of ± 8 and 16 in case of ± 16
Step 2:	Initialize the harmony memory with five individuals ($HMS = 5$), where each decision variable u and v of the candidate motion vector MV_a is set according to the fixed pattern shown in Fig. 2.6. Considering $a \in (1, 2, \ldots, HMS)$. Define also the individual database array $\mathbf{T}$, as an empty array
Step 3:	Compute the fitness values for each individual according to the fitness calculation strategy presented in Sect. 2.3. Since all individuals of the initial population fulfill rule 2 conditions, they are evaluated through a real fitness function (calculating the real SAD values)
Step 4:	Update the new evaluations in the individual database array $\mathbf{T}$
Step 5:	Determine the candidate solution MV_w of HMS holding the worst fitness value

Step 6:	Improvise a new harmony MV_{new} such that:
	 for ($j = 1$ to 2) do
	    if ($r_1 < HMCR$) then
	        $MV_{new}(j) = MV_a(j)$ where a is element of $(1, 2, \ldots, HMS)$
	        randomly selected
	        if ($r_2 < PAR$) then
	        $MV_{new}(j) = MV_{new}(j) \pm r_3 \cdot BW$ where
	        $r_1, r_2, r_3 \in (0, 1)$
	        if $MV_{new}(j) < l(j)$
	        $MV_{new}(j) = l(j)$
	        end if
	        if $MV_{new}(j) > u(j)$
	        $MV_{new}(j) = u(j)$
	        end if
	        end if
	    else
	        $MV_{new}(j) = 1 + \mathrm{round}(r \cdot E_p)$, where $r \in (-1, 1)$, $E_p = 8$ or 16
	    end if
	 end for
	 $E_p = 8$, in case of a search window of ± 8 and 16 in case of ± 16

Step 7:	Compute the fitness value of MV_{new} by using the fitness calculation strategy presented in Sect. 2.3
Step 8:	Update the new evaluation in the individual database array $\mathbf{T}$
Step 9:	Update HM. In case that the fitness value (evaluated or approximated) of the new solution MV_{new}, is better than the solution MV_w, such position is selected as an element of HM, otherwise the solution MV_w remains
Step 10:	Determine the best individual of the current new population. If the new fitness (SAD) value is better than the old best fitness value, then update $\hat{u}_{best}$, $\hat{v}_{best}$
Step 11:	If the number of iterations (NI) has been reached (25 in the case of a search window of ± 8 and 45 for ± 16), then the MV is $\hat{u}_{best}$, $\hat{v}_{best}$; otherwise go back to Step 5

Thus, the presented HS-BM algorithm considers different search locations, 30 in the case of a search window of ± 8 and 50 for ± 16, during the complete optimization process (which consists of 25 and 45 different iterations depending on the search window, plus the five initial positions). However, only a few search locations are evaluated using the actual fitness function (5–14, in the case of a search window of ±8 and 7–22, for ±16) while the remaining positions are just estimated. Therefore, as the evaluated individuals and their respective fitness values are exclusively stored in the array **T**, the resources used for the management of such data are negligible. Figure 2.7 shows two search-patterns examples that have been generated by the HS-BM approach. Such patterns exhibit the evaluated search-locations (rule 1 and 2) in white-cells, whereas the minimum location is marked in black. Grey-cells represent cells that have been estimated (rule 3) or not visited during the optimization process.

2.5.4 Discussion on the Accuracy of the Fitness Approximation Strategy

HS has been found to be capable of solving several practical optimization problems. A distinguishing feature of HS is about its operation with only a population of individuals. It uses multiple candidate solutions at each step. This requires the computation of the fitness function for each candidate at every iteration. The ability to locate the global optimum depends on sufficient exploration of the search space which requires the use of enough individuals. Under such circumstances, this work intends to couple the HS method with a fitness approximation model in order to replace (when it is feasible) the use of an expensive fitness function to compute the quality of several individuals.

Similar to other EA approaches, HS maintains two different phases on its operation: exploitation and exploration [58]. Exploitation (local search) refers to the

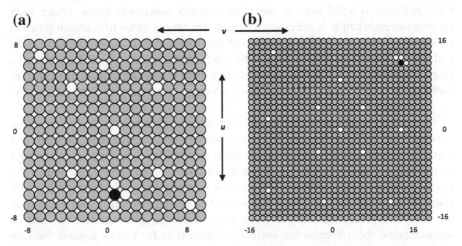

Fig. 2.7 Search-patterns generated by the HS-BM algorithm. **a** Search window pattern ±8 with solution $\hat{u}_{best}^1 = 0$ and $\hat{v}_{best}^1 = -6$. **b** Search window pattern ±16 with solution $\hat{u}_{best}^2 = 11$ and $\hat{v}_{best}^2 = 12$

action of refining the best solutions found so far whereas exploration (global search) represents the process of generating semi-random individuals in order to capture information of unexplored areas. In spite of this, the optimization process is guided by the best individuals seen-so-far [59]. They are selected more frequently and thereby modified or combined by the evolutionary operators in order to generate new promising individuals.

Therefore, the main concern in using fitness approximation models, is to accurately calculate the quality of those individuals which either hold great possibilities of grasping an excellent fitness value (individuals that are to close to one of the best individuals seen-so-far), or do not have reference about their possible fitness values (individuals located in unexplored areas) [60, 61]. Most of the fitness approximation methods proposed in the literature [48–50] use interpolation models in order to compute the fitness value of new individuals. Since the estimated fitness value is approximated considering other individuals which might be located far away from the position to be calculated, it introduces big errors that harshly affects the optimization procedure [45]. Different to such methods, in our approach, the fitness values are calculated using three different rules which promote the evaluation of individuals that require particular accuracy (Rule 1 and Rule 2). On the other hand, the strategy estimates those individuals which according to the evidence known so-far (elements contained in the array T) represent unacceptable solutions (bad fitness values). Such individuals do not play an important roll in the optimization process, therefore their accuracy is not considered critical [46, 62].

It is important to emphasize that the presented fitness approximation strategy has been designed considering some of the BM process particularities. Error landscapes in BM, due to the continuity principle [17, 20, 63] of video sequences, present the following particularity: the closer neighbors to one global/local minimum (a motion vector with a low SAD value) decrement their SAD value as they approach to it. Such behavior is valid even in the most complex movement types. Under such circumstances, when it is necessary calculate the fitness value of a search position which is close to one of the search position previously visited (according to array T) and whose fitness value was unacceptable, its fitness value is estimated according to rule 3. This decision is taken considering that there is a strong evidence to consider such position as a bad individual from which it is no necessary to get a good accuracy level.

Figure 2.8 presents the optimization procedure achieved by the combination between HS and the fitness approximation strategy presented in this chapter over a complex movement case. The example illustrates the fitness strategy operation for a complex movement considering a search window of ± 8. Figure 2.8a shows the error landscape (SAD values) in a 3-D view, whereas Fig. 2.8b depicts the search positions calculated by the fitness approximation strategy over the SAD values that are computed for all elements of the search window as reference (for the sake of representation, both Figures are normalized from 0 to 1). Yellow squares indicate evaluated search positions whereas blue squares represent the estimated ones. Since random numbers are involved by HS in the generation of new individuals, they may encounter same positions (repetition) that other individuals have visited in previous

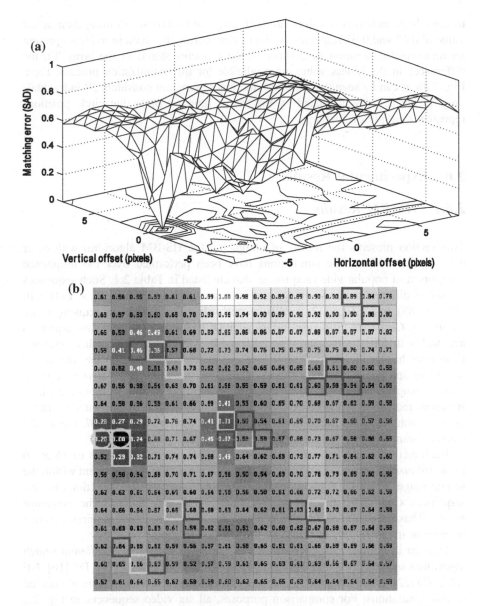

Fig. 2.8 Example of the optimization procedure: **a** Error landscape (SAD values) in a 3-D view. **b** Search positions calculated by the fitness approximation strategy over the SAD values which are computed for all elements of the search window of size ±8

iterations. Circles represent search positions that have been selected several times during the optimization procedure. The problem of accuracy, in the estimation process, can also be appreciated through a close analysis from the red dashed square of Fig. 2.8b. As the blue squares represent the estimated search positions according to the rule 3, their fitness values are both assigned to 0.68 substituting their actual value of 0.63 and 0.67 respectively. Thus, considering that such individuals present an unacceptable solution (according to the elements stored in the array **T**), the differences in the fitness value are negligible for the optimization process. From Fig. 2.8b, it can be seen that although the fitness function considers 30 individuals only 12 are actually evaluated by the fitness function (note that circle positions represent multiple evaluations).

2.6 Experimental Results

2.6.1 HS-BM Results

This section presents the results of comparing the HS-BM algorithm with other existing fast BMAs. The simulations have been performed over the luminance component of popular video sequences that are listed in Table 2.1. Such sequences consist of different degrees and types of motion including QCIF (176 × 144), CIF (352 × 288) and SIF (352 × 240) respectively. The first four sequences are *Container*, *Carphone*, *Foreman* and *Akiyo* in QCIF format. The next two sequences are *Stefan* in CIF format and *Football* in SIF format. Among such sequences, *Container* has gentle, smooth and low motion changes and consists mainly of stationary and quasi-stationary blocks. *Carphone*, *Foreman* and *Akiyo* have moderately complex motion getting a "medium" category regarding its motion content. Rigorous motion which is based on camera panning with translation and complex motion content can be found in the sequences of *Stefan* and *Football*. Figure 2.9 shows a sample frame from each sequence.

Each picture frame is partitioned into macro-blocks with the sizes of 16 × 16 ($N = 16$) pixels for motion estimation, where the maximum displacement within the search range W is of ±8 pixels in both horizontal and vertical directions for the sequences *Container*, *Carphone*, *Foreman*, *Akiyo* and Stefan. The sequence *Football* has been simulated with a window size W of ±16, which requires a greater number of iterations (8 iterations) by the HS-BM method.

In order to compare the performance of the HS-BM approach, different search algorithms such as FSA, TSS [12], 4SS [15], NTSS [13], BBGD [19], DS [16], NE [20], ND [22], LWG [29], GFSS [30] and PSO-BM [31] have been all implemented in our simulations. For comparison purposes, all six video sequences in Fig. 2.8 have been all used. All simulations are performed on a Pentium IV 3.2 GHz PC with 1 GB of memory.

In the comparison, two relevant performance indexes have been considered: the distortion performance and the search efficiency.

Distortion performance

First, all algorithms are compared in terms of their distortion performance which is characterized by the peak-signal-to-noise-ratio (PSNR) value. Such value

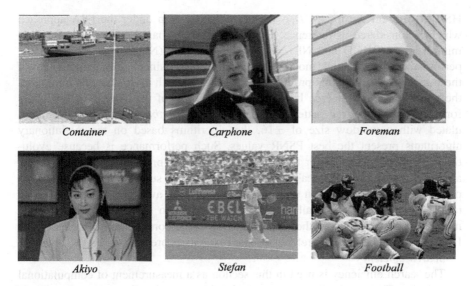

Fig. 2.9 Test video sequences

indicates the reconstruction quality when motion vectors, which are computed through a BM approach, are used. In PSNR, the signal is the original data frames whereas the noise is the error introduced by the calculated motion vectors. The PSNR is thus defined as:

$$PSNR = 10 \times \log_{10}\left(\frac{255^2}{MSE}\right) \quad (2.7)$$

where *MSE* is the mean square between the original frames and those compensated by the motion vectors. Additionally, as an alternative performance index, it is used in the comparison the PSNR degradation ratio (D_{PSNR}). This ratio expresses in percentage (%) the level of mismatch between the PSNR of a BM approach and the PSNR of the FSA which is considered as reference. Thus the D_{PSNR} is defined as

$$D_{PSNR} = -\left(\frac{PSNR_{FSA} - PSNR_{BM}}{PSNR_{FSA}}\right) \times 100\% \quad (2.8)$$

Table 2.2 shows the comparison of the PSNR values and the PSNR degradation ratios (D_{PSNR}) among the BM algorithms. The experiment considers the six image sequences presented in Fig. 2.8. As it can be seen, in the case of the slow-moving sequence *Container*, the PSNR values (the D_{PSNR} ratios) of all BM algorithms are similar. For the medium motion content sequences such as *Carphone*, *Foreman* and *Akiyo*, the approaches consistent of fixed patterns (TSS, 4SS and NTSS) exhibit the worst PSNR value (high D_{PSNR} ratio) except for the DS algorithm. On the other hand, BM methods that use evolutionary algorithms (LWG, GFSS, PSO-BM and

HS-BM) present the lowest D_{PSNR} ratio, only one step under the FSA approach which is considered as reference. Finally, approaches based on the error-function minimization (BBGD and NE) and pixel-decimation (ND), show an acceptable performance. For the high motion sequence of *Stefan*, since the motion content of these sequences is complex producing error surfaces with more than one minimum, the performance, in general, becomes worst for most of the algorithms especially for those based on fixed patterns. In the sequence *Football*, which has been simulated with a window size of ±16, the algorithms based on the evolutionary algorithms present the best PSNR values. Such performance is because evolutionary methods adapt better to complex optimization problems where the search area and the number of local minima increase. As a summary of the distortion performance, the last column of Table 2.2 presents the average PSNR degradation ratio (D_{PSNR}) obtained for all sequences. According to such values, the HS-BM method is superior to any other approach. Due to the computation complexity, the FSA is considered just as a reference. The best entries are bold-cased in Table 2.2.

Search efficiency

The search efficiency is used in this section as a measurement of computational complexity. The search efficiency is calculated by counting the average number of search points (or the average number of SAD computations) for the MV estimation. In Table 2.3, the search efficiency is compared, where the best entries are bold-cased. Just above FSA, some evolutionary algorithms such as LWG, GFSS and PSO-BM hold the highest number of search points per block. On the contrary, the HS-BM algorithm can be considered as a fast approach as it maintains a similar performance to DS. From data shown in Table 2.3, the average number of search locations, corresponding to the HS-BM method, represents the number of SAD evaluations (the number of SAD estimations are not considered whatsoever). Additionally, the last two columns of Table 2.3 present the number of search locations that have been averaged (over the six sequences) and their performance rank. According to these values, the HS-BM method is ranked in the first place. The average number of search points visited by the HS-BM algorithm ranges from 9.2 to 17.3, representing the 4% and the 7.4% respectively in comparison to the FSA method. Such results demonstrate that our approach can significantly reduce the number of search points. Hence, the HS-BM algorithm presented in this chapter is comparable to the fastest algorithms and delivers a similar precision to the FSA approach.

2.6.2   Results on H.264

Other set of experiments have been performed in JM-12.2 [64] of H.264/AVC reference software. In the simulations, we compare FS, DS [16], EPZS [24], TSS [12], 4SS [15], NTSS [13], BBGD [19] and the HS-BM algorithm in terms of coding efficiency and computational complexity.

Table 2.2 PSNR values and D_{PSNR} comparison of the BM methods

Algorithm	Container $W = \pm 8$		Carphone $W = \pm 8$		Foreman $W = \pm 8$		Akiyo $W = \pm 8$		Stefan $W = \pm 8$		Football $W = \pm 16$		Total average (D_{PSNR})
	PSNR	D_{PSNR}	PSNR	D_{PSNR}	PSNR	D_{PSNR}	PSNR	D_{PSNR}	PSNR	D_{PSNR}	PSNR	D_{PSNR}	
FSA	43.18	0	31.51	0	31.69	0	29.07	0	25.95	0	23.07	0	0
TSS	43.10	−0.20	30.27	−3.92	29.37	−7.32	26.21	−9.84	21.14	−18.52	20.03	−13.17	−8.82
4SS	43.12	−0.15	30.24	−4.01	29.34	−7.44	26.21	−9.84	21.41	−17.48	20.10	−12.87	−8.63
NTSS	43.12	−0.15	30.35	−3.67	30.56	−3.57	27.12	−6.71	22.52	−13.20	20.21	−12.39	−6.61
BBGD	43.14	−0.11	31.30	−0.67	31.00	−2.19	28.10	−3.33	25.17	−3.01	22.03	−4.33	−2.27
DS	43.13	−0.13	31.26	−0.79	31.19	−1.59	28.00	−3.70	24.98	−3.73	22.35	−3.12	−2.17
NE	43.15	−0.08	31.36	−0.47	31.23	−1.47	28.53	−2.69	25.22	−2.81	22.66	−1.77	−1.54
ND	43.15	−0.08	31.35	−0.50	31.20	−1.54	28.32	−2.56	25.21	−2.86	22.60	−2.03	−1.59
LWG	43.16	−0.06	31.40	−0.36	31.31	−1.21	28.71	−1.22	25.41	−2.09	22.90	−0.73	−0.95
GFSS	43.15	−0.06	31.38	−0.40	31.29	−1.26	28.69	−1.28	25.34	−2.36	22.92	−0.65	−1.01
PSO-BM	43.15	−0.07	31.39	−0.38	31.27	−1.34	28.65	1.42	25.39	−2.15	22.88	−0.82	−1.03
HS-BM	43.16	−0.03	31.49	−0.03	31.63	−0.21	29.01	−0.18	25.89	−0.20	23.01	−0.20	−0.18

Table 2.3 Averaged number of visited search points per block for all ten BM methods

Algorithm	Container $W = \pm 8$	Carphone $W = \pm 8$	Foreman $W = \pm 8$	Akiyo $W = \pm 8$	Stefan $W = \pm 8$	Football $W = \pm 16$	Total average	Rank
FSA	289	289	289	289	289	1089	422.3	12
TSS	25	25	25	25	25	25	25	8
4SS	19	25.5	24.8	27.3	25.3	25.6	24.58	7
NTSS	17.2	21.8	22.1	23.5	25.4	26.5	22.75	6
BBGD	9.1	14.5	14.5	13.2	17.2	22.3	15.13	3
DS	7.5	12.5	13.4	11.8	15.2	17.8	13.15	2
NE	11.7	13.8	14.2	14.5	19.2	24.2	16.36	5
ND	10.8	13.4	13.8	14.1	18.4	25.1	16.01	4
LWG	75	75	75	75	75	75	75	11
GFSS	60	60	60	60	60	60	60	10
PSO-BM	32.5	48.5	48.1	48.5	52.2	52.2	47	9
HS-BM	8.0	12.2	11.2	11.5	17.1	15.2	12.50	1

For encoding purposes JM-12.2 Main Encoder Profile has been used. For each test sequence only the first frame has been coded as I frame and the remaining frames are coded as P frames. Only one reference frame has been used. Each pixel in the image sequences is uniformly quantized to 8 bits. Sum of absolute difference (SAD) distortion function is used as the block distortion measure (BDM). Image formats used are QCIF, CIF and SIF meanwhile sequences are tested at 30 fps (frames per second). The simulation platform in our experiments is a PC with Intel Pentium IV 2.66 GHz CPU.

The test sequences used for our experiments are *Container*, *Akiyo* and *Football*. These sequences exhibit a variety of motion that is generally encountered in real video. For the sequences *Container* and *Akiyo* a search window of ±8 is selected meanwhile for the football sequence a search window of ±16 is considered. The group of experiments has been performed over such sequences at four different quantization parameters (QP = 28, 32, 36, 40) in order to test the algorithms at different transmission conditions.

(a) *Coding efficiency*

In the first experiment, the performance of the HS-BM algorithm is compared to other BM algorithms regarding the coding efficiency. Two different performance indexes are used for evaluating the coding quality: the PSNR Gain and the increasing of the Bit Rate. In order to comparatively assess the results, two additional indexes, called PSNR loss and Bit Rate Incr., relate the performance of each method with the FSA performance as a reference. Such indexes are calculated as follows:

$$\text{PSNR loss} = \text{PSNR FSA} - \text{PSNR algorithm} \qquad (2.9)$$

$$\text{Bit Rate Incr.} = \left(\frac{\text{Bit Rate algorithm} - \text{Bit Rate FSA}}{\text{Bit Rate FSA}} \right) \times 100 \qquad (2.10)$$

Table 2.4 Coding efficiency results for the *container* sequence, considering a window size W of ±8

BM	Coding efficiency				Computational complexity		
	PSNR	Bit-rate (Kbits/s)	PNSR loss (dB)	Bit-rate increase (%)	ACT (ms)	IN	CM (Bytes)
FSA	36.06	41.4	–	–	133.2	122	3072
DS	36.04	43.4	0.02	4.83	6.33	138	420
EPZS	36.04	41.3	0.02	−0.20	19.5	621	8972
TSS	34.01	45.2	2.05	9.17	2.1	100	180
4SS	35.22	44.7	0.84	7.97	2.8	100	204
NTSS	35.76	44.3	0.30	7.00	3.7	110	256
BBGD	35.98	42.1	0.08	1.70	9.1	256	1024
HS-BM	36.04	41.5	0.02	0.20	3.8	189	784

Table 2.5 Simulation results for the *Akiyo* sequence, considering a window size W of ± 8

BM	Coding efficiency				Computational complexity		
	PSNR	Bit-rate (Kbits/s)	PNSR loss (dB)	Bit-rate increase (%)	ACT (ms)	IN	CM (Bytes)
FSA	38.19	25.6	–	–	133.2	122	3072
DS	38.11	25.9	0.08	1.20	7.45	138	420
EPZS	38.19	25.3	–	−1.20	22.1	621	8972
TSS	30.32	29.3	7.87	14.45	2.1	100	180
4SS	32.42	28.4	5.77	10.93	2.8	100	204
NTSS	33.57	27.2	4.62	6.25	3.7	110	256
BBGD	35.21	26.8	2.98	4.68	10.1	256	1024
HS-BM	38.17	25.5	0.02	−0.40	3.9	189	784

Table 2.6 Simulation results for the *Football* sequence, considering a window size W of ± 16

BM	Coding efficiency				Computational complexity		
	PSNR	Bit-rate (Kbits/s)	PNSR loss (dB)	Bit-rate increase (%)	ACT (ms)	IN	CM (Bytes)
FSA	34.74	98.85	–	–	245.7	122	12,288
DS	32.22	99.89	2.52	1.02	10.36	144	600
EPZS	34.72	98.81	0.02	−0.04	26.8	678	20,256
TSS	27.12	106.42	7.62	7.65	2.9	113	180
4SS	27.91	105.29	6.83	6.51	3.1	113	204
NTSS	29.11	104.87	5.63	6.09	4.2	113	256
BBGD	29.76	103.96	4.98	5.19	16.41	268	2048
HS-BM	34.73	98.91	0.01	0.06	4.1	201	1024

Tables 2.4, 2.5 and 2.6 show a coding efficiency comparison among BM algorithms. It is observed, from experimental results, that the HS-BM algorithm holds an effective coding quality because the loss in terms of PSNR and the increase of the Bit rate are low with an average of 1.6 dB and −0.04%, respectively. Such coding performance is similar to the one produced by the EPZS method whereas it is much better than the obtained by other BM algorithms which posses the worst coding quality.

(b) *Computational complexity*

In the second experiment, we have compared the performance of the HS-BM algorithm to other BM algorithms in terms of computational overhead. As the JM-12.2 platform allows to simulate BM algorithms in real time conditions, we have used such results in order to evaluate their performances.

Three different performance indexes are used for evaluating the computational complexity; they are the averaged coding time (ACT), instruction number (IN) and consumed memory (CM). The ACT is the averaged time employed to codify a complete frame (the averaged time consumed after finding all the corresponding motion vectors for a frame). IN represents the number of instructions used to implement each algorithm in the JM-12.2 profile. CM considers the memory size used by the JM-12.2 platform in order to manage the data that are employed by each BM algorithm.

Tables 2.4, 2.5 and 2.6 show the computational complexity comparison among the BM algorithms. It is observed from the experimental results that the HS-BM algorithm possesses a competitive ACT value (from 3.8 to 4.1 ms) in comparison to other BM algorithms. This fact reflexes that although the cost of applying the fitness approximation strategy represents an overhead that is not required in most fast BM methods, such overhead is negligible in comparison to the cost of the number of fitness evaluations which have been saved. The ACT values, presented by the HS-BM, are lightly superior to those produced by the fast BM methods (TSS, 4SS and NTSS) whereas it is much better than those generated by the EPZS algorithm which possesses the worst computational performance. On the other hand, the resources (in terms of number of instructions IN and required memory CM) needed by the HS-BM approach are considered as standard in software and hardware architectures.

2.7 Conclusions

In this chapter, a block-matching algorithm that combines harmony search with a fitness approximation model is presented. The approach uses as potential solutions the motion vectors belonging to the search window. A fitness function evaluates the matching quality of each motion vector candidate. To save computational time, the approach incorporates a fitness calculation strategy to decide which motion vectors can be estimated or actually evaluated. Guided by the values given by such fitness calculation strategy, the set of motion vectors are evolved using the HS operators so the best possible motion vector can be identify.

Since the presented algorithm does not consider any fixed search pattern during the BM process or any other movement assumption, a high probability for finding the true minimum (accurate motion vector) is expected regardless of the movement complexity contained in the sequence. Therefore, the chance of being trapped into a local minimum is reduced in comparison to other BM algorithms.

The performance of HS-BM has been compared to other existing BM algorithms by considering different sequences which present a great variety of formats and movement types. Experimental results demonstrate that the presented algorithm maintains the best balance between coding efficiency and computational complexity.

Although the experimental results indicate that the HS-BM method can yield better results on complicated sequences, it should be noticed that the aim of this chapter is to show that the fitness approximation can effectively serve as an attractive alternative to evolutionary algorithms for solving complex optimization problems, yet demanding fewer function evaluations.

References

1. Cirrincione G, Cirrincione M (2003) A novel self-organizing neural network for motion segmentation. Appl Intell 18(1):27–35
2. Risinger L, Kaikhah K (2008) Motion detection and object tracking with discrete leaky integrate-and-fire neurons. Appl Intell 29(3):248–262
3. Bohlooli A, Jamshidi K (2012) A GPS-free method for vehicle future movement directions prediction using SOM for VANET. Appl Intell 36(3):685–697
4. Kang J-G, Kim S, An S-Y, Se-Young O (2012) A new approach to simultaneous localization and map building with implicit model learning using neuro evolutionary optimization. Appl Intell 36(1):242–269
5. Tzovaras D, Kompatsiaris I, Strintzis MG (1999) 3D object articulation and motion estimation in model-based stereoscopic videoconference image sequence analysis and coding. Sig Process Image Commun 14(10):817–840
6. Barron JL, Fleet DJ, Beauchemin SS (1994) Performance of optical flow techniques. Int J Comput Vis 12(1):43–77
7. Skowronski J (1999) Pel recursive motion estimation and compensation in subbands. Sig Process Image Commun 14:389–396
8. Huang T, Chen C, Tsai C, Shen C, Chen L (2006) Survey on block matching motion estimation algorithms and architectures with new results. J VLSI Sig Proc 42:297–320
9. MPEG4 (2000) Information technology coding of audio visual objects part 2: visual, JTC1/SC29/WG11, ISO/IEC14469-2(MPEG-4Visual)
10. H.264 (2003) Joint Video Team (JVT) of ITU-T and ISO/IEC JTC1, Geneva, JVT ofISO/IEC MPEG and ITU-T VCEG, JVT-g050r1, Draft ITU-TRec. and Final Draft International Standard of Joint Video Specification (ITU-T Rec.H.264-ISO/IEC14496-10AVC)
11. Jain JR, Jain AK (1981) Displacement measurement and its application in interframe image coding. IEEE Trans Commun COM-29:1799–1808
12. Jong H-M, Chen L-G, Chiueh T-D (1994) Accuracy improvement and cost reduction of 3-step search block matching algorithm for video coding. IEEE Trans Circ Syst Video Technol 4:88–90
13. Li R, Zeng B, Liou ML (1994) A new three-step search algorithm for block motion estimation. IEEE Trans Circ Syst Video Technol 4(4):438–442
14. Jianhua L, Liou ML (1997) A simple and efficient search algorithm for block-matching motion estimation. IEEE Trans Circ Syst Video Technol 7(2):429–433
15. Po L-M, Ma W-C (1996) A novel four-step search algorithm for fast block motion estimation. IEEE Trans Circ Syst Video Technol 6(3):313–317
16. Zhu S, Ma K-K (2000) A new diamond search algorithm for fast block-matching motion estimation. IEEE Trans Image Process 9(2):287–290
17. Nie Y, Ma K-K (2002) Adaptive rood pattern search for fast block-matching motion estimation. IEEE Trans Image Process 11(12):1442–1448
18. Yi-Ching L, Jim L, Zuu-Chang H (2009) Fast block matching using prediction and rejection criteria. Sig Process 89:1115–1120
19. Liu L, Feig E (1996) A block-based gradient descent search algorithm for block motion estimation in video coding. IEEE Trans Circ Syst Video Technol 6(4):419–422
20. Saha A, Mukherjee J, Sural S (2011) A neighborhood elimination approach for block matching in motion estimation. Sig Process Image Commun 26(8–9):438–454
21. Chow KHK, Liou ML (1993) Generic motion search algorithm for video compression. IEEE Trans Circ Syst Video Technol 3:440–445
22. Saha A, Mukherjee J, Sural S (2008) New pixel-decimation patterns for block matching in motion estimation. Sig Process Image Commun 23:725–738
23. Song Y, Ikenaga T, Goto S (2007) Lossy strict multilevel successive elimination algorithm for fast motion estimation. IEICE Trans Fundam E90(4):764–770

24. Tourapis AM (2002) Enhanced predictive zonal search for single and multiple frame motion estimation. In: Proceedings of visual communications and image processing, California, pp 1069–1079

25. Chen Z, Zhou P, He Y, Chen Y (2002) Fast integer pel and fractional pel motion estimation for JVT, ITU-T. Doc. #JVT-F-017

26. Nisar H, Malik AS, Choi T-S (2012) Content adaptive fast motion estimation based on spatio-temporal homogeneity analysis and motion classification. Pattern Recogn Lett 33:52–61

27. Holland JH (1975) Adaptation in natural and artificial systems. University of Michigan Press, Ann Arbor

28. Kennedy J, Eberhart RC (1995) Particle swarm optimization. In: Proceedings of the 1995 IEEE International conference on neural networks, vol 4, pp 1942–1948

29. Chun-Hung L, Ja-Ling W (1998) A lightweight genetic block-matching algorithm for video coding. IEEE Trans Circ Syst Video Technol 8(4):386–392

30. Wu A, So S (2003) VLSI implementation of genetic four-step search for block matching algorithm. IEEE Trans Consum Electron 49(4):1474–1481

31. Yuan X, Shen X (2008) Block matching algorithm based on particle swarm optimization. In: International conference on embedded software and systems (ICESS 2008), Sichuan

32. Geem ZW, Kim JH, Loganathan GV (2001) A new heuristic optimization algorithm: harmony search. Simulations 76:60–68

33. Mahdavi M, Fesanghary M, Damangir E (2007) An improved harmony search algorithm for solving optimization problems. Appl Math Comput 188:1567–1579

34. Omran MGH, Mahdavi M (2008) Global-best harmony search. Appl Math Comput 198:643–656

35. Lee KS, Geem ZW (2005) A new meta-heuristic algorithm for continuous engineering optimization, harmony search theory and practice. Comput Methods Appl Mech Eng 194:3902–3933

36. Lee KS, Geem ZW, Lee SH, Bae K-W (2005) The harmony search heuristic algorithm for discrete structural optimization. Eng Optim 37:663–684

37. Kim JH, Geem ZW, Kim ES (2001) Parameter estimation of the nonlinear Muskingum model using harmony search. J Am Water Resour Assoc 37:1131–1138

38. Geem ZW (2006) Optimal cost design of water distribution networks using harmony search. Eng Optim 38:259–280

39. Lee KS, Geem ZW (2004) A new structural optimization method based on the harmony search algorithm. Comput Struct 82:781–798

40. Ayvaz TM (2007) Simultaneous determination of aquifer parameters and zone structures with fuzzy c-means clustering and meta-heuristic harmony search algorithm. Adv Water Resour 30:2326–2338

41. Geem ZW, Lee KS, Park YJ (2005) Application of harmony search to vehicle routing. Am J Appl Sci 2:1552–1557

42. Geem ZW (2008) Novel derivative of harmony search algorithm for discrete design variables. Appl Math Comput 199(1):223–230

43. Cuevas E, Ortega-Sánchez N, Zaldivar D, Pérez-Cisneros M (2012) Circle detection by harmony search optimization. J Intell Rob Syst 66(3):359–376

44. Jin Y (2005) Comprehensive survey of fitness approximation in evolutionary computation. Soft Comput 9:3–12

45. Jin Yaochu (2011) Surrogate-assisted evolutionary computation: recent advances and future challenges. Swarm Evolut Comput 1:61–70

46. Branke J, Schmidt C (2005) Faster convergence by means of fitness estimation. Soft Comput 9:13–20

47. Zhou Z, Ong Y, Nguyen M, Lim D (2005) A study on polynomial regression and gaussian process global surrogate model in hierarchical surrogate-assisted evolutionary algorithm. In: IEEE congress on evolutionary computation (ECiDUE'05), Edinburgh, 2–5 Sept 2005

48. Ratle A (2001) Kriging as a surrogate fitness landscape in evolutionary optimization. Artif Intell Eng Des Anal Manuf 15:37–49
49. Lim D, Jin Y, Ong Y, Sendhoff B (2010) Generalizing surrogate-assisted evolutionary computation. IEEE Trans Evol Comput 14(3):329–355
50. Ong Y, Lum K, Nair P (2008) Evolutionary algorithm with hermite radial basis function interpolants for computationally expensive adjoint solvers. Comput Optim Appl 39(1):97–119
51. Luoa C, Shao-Liang Z, Wanga C, Jiang Z (2011) A metamodel-assisted evolutionary algorithm for expensive optimization. J Comput Appl Math. doi:10.1016/j.cam.2011.05.047
52. Goldberg DE (1989) Genetic algorithms in search, optimization and machine learning. Addison-Wesley Professional, Menlo Park
53. Li X, Xiao N, Claramunt C, Lin H (2011) Initialization strategies to enhancing the performance of genetic algorithms for the p-median problem. Comput Ind Eng. doi:10.1016/j. cie.2011.06.015
54. Xiao N (2008) A unified conceptual framework for geographical optimization using evolutionary algorithms. Ann Assoc Am Geogr 98:795–817
55. Soak S-M, Lee S-W (2012) A memetic algorithm for the quadratic multiple container packing problem. Appl Intell 36(1):119–135
56. Luque C, Valls JM, Isasi P (2011) Time series prediction evolving Voronoi regions. Appl Intell 34(1):116–126
57. Montero E, Riff M-C (2011) On-the-fly calibrating strategies for evolutionary algorithms. Inf Sci 181:552–566
58. Tan KC, Chiam SC, Mamun AA, Goh CK (2009) Balancing exploration and exploitation with adaptive variation for evolutionary multi-objective optimization. Eur J Oper Res 197 (2):701–713
59. Wang P, Zhang J, Xub L, Wang H, Feng S, Zhu H (2011) How to measure adaptation complexity in evolvable systems—a new synthetic approach of constructing fitness functions. Expert Syst Appl 38:10414–10419
60. Tenne Y (2012) A computational intelligence algorithm for expensive engineering optimization problems. Eng Appl Artif Intell 25(5):1009–1021
61. Büche D, Schraudolph NN, Koumoutsakos P (2005) Accelerating evolutionary algorithms with Gaussian process fitness function models. IEEE Trans Syst Man Cybern Part C Appl Rev 35(2):183–194
62. Giannakoglou KC, Papadimitriou DI, Kampolis IC (2006) Aerodynamic shape design using evolutionary algorithms and new gradient-assisted metamodels. Comput Methods Appl Mech Eng 195:6312–6329
63. Tai S-C, Chen Y-R, Chen Y-H (2007) Small-diamond-based search algorithm for fast block motion estimation. Sig Process Image Commun 22:877–890
64. Joint Video Team Reference Software (2007) Version 12.2 (JM12.2). http://iphome.hhi.de/ suehring/tml/download/

Chapter 3
Gravitational Search Algorithm Applied to Parameter Identification for Induction Motors

Induction motors represent the main component in most of the industries. Induction motors represent the main component in most of the industries. They use the biggest energy percentages in industrial facilities. This consume depends on the operation conditions of the induction motor imposed by its internal parameters. In this approach, the parameter estimation process is transformed into a multidimensional optimization problem where the internal parameters of the induction motor are considered as decision variables. Thus, the complexity of the optimization problem tends to produce multimodal error surfaces in which their cost functions are significantly difficult to minimize. Several algorithms based on evolutionary computation principles have been successfully applied to identify the optimal parameters of induction motors. However, most of them frequently acquire sub-optimal solutions as a result of an inappropriate balance between exploitation and exploration in their search strategies. This chapter presents an algorithm for the optimal parameter identification of induction motors that uses the recent evolutionary method called the Gravitational Search Algorithm (GSA). In general, GSA presents a better performance in multimodal problems, avoiding critical flaws such as the premature convergence to sub-optimal solutions. The presented algorithm has been tested on several models and its simulation results show the effectiveness of the scheme.

3.1 Introduction

The environmental consequences that overconsumption of electrical energy entails has recently attracted the attention in different fields of the engineering. Therefore, the improvement of machinery and elements that have high electrical energy consumption have become an important task nowadays [1].

Induction motors present several benefits such as their ruggedness, low price, cheap maintenance and easy controlling [2]. However, more than a half of electric

© Springer International Publishing AG 2017   45
M.-A. Díaz-Cortés et al., *Engineering Applications of Soft Computing*,
Intelligent Systems Reference Library 129, DOI 10.1007/978-3-319-57813-2_3

energy consumed by industrial facilities is due to use of induction motors. With the massive use of induction motors, the electrical energy consumption has increased exponentially through years. This fact has generated the need to improve their efficiency which mainly depends on their internal parameters. The parameter identification of induction motors represents a complex task due to its non-linearity. As a consequence, different alternatives have been proposed in the literature. Some examples include the proposed by Waters and Willoughby [3], where the parameter are estimated from the knowledge of certain variables such as stator resistance and the leakage reactance, the proposed by Ansuj et al. [4], where the identification is based on a sensitivity analysis and the proposed by De Kock et al. [5], where the estimation is conducted through an output error technique.

As an alternative to such techniques, the problem of parameter estimation in induction motors has also been handled through evolutionary methods. In general, they have demonstrated, under several circumstances, to deliver better results than those based on deterministic approaches in terms of accuracy and robustness [6]. Some examples of these approaches used in the identification of parameters in induction motors involve methods such as genetic algorithms (GA) [7], particle swarm optimization (PSO) [8, 9], artificial immune system (AIS) [10], bacterial foraging algorithm (BFA) [11], shuffled frog-leaping algorithm [12], hybrid of genetic algorithm and PSO [6], multiple-global-best guided artificial bee colony [13], just to mention a few. Although these algorithms present interesting results, they have an important limitation: They frequently obtain sub-optimal solutions as a consequence of the limited balance between exploration and exploitation in their search strategies.

On the other hand, the gravitational search algorithm (GSA) [14] is a recent evolutionary computation algorithm which is inspired on physical phenomenon of the gravity. In GSA, its evolutionary operators are built considering the principles of the gravitation. Different to the most of existent evolutionary algorithms, GSA presents a better performance in multimodal problems, avoiding critical flaws such as the premature convergence to sub-optimal solutions [15, 16]. Such characteristics have motivated the use of SSO to solve an extensive variety of engineering applications such as energy [17], image processing [18] and machine learning [19].

This chapter describes an algorithm for the optimal parameter identification of induction motors. To determine the parameters, the presented method uses a recent evolutionary method called the GSA. A comparison with state-of-the-art methods such as artificial bee colony (ABC) [20], differential evolution (DE) [21] and PSO [22] on different induction models has been incorporated to demonstrate the performance of the presented approach. Conclusions of the experimental comparison are validated through statistical tests that properly support the discussion.

The sections of this chapter are organized as follows: in Sect. 3.2 problem statements are presented, Sect. 3.3 describes the evolutionary technique used GSA, Sect. 3.4 shows experimental results considering the comparison with DE, ABC and PSO and a non-parametric statistical validation and finally in Sect. 3.5 conclusions are discussed.

3.2 Problem Statement

An induction motor can be represented as a steady-state equivalent circuit and treated as a least square optimization problem, which, due its nature highly non-linear it becomes difficult its minimization. The main objective is minimizing the error between the calculated and the manufacturer data adjusting the parameters of an induction motor equivalent circuit. In this chapter we use the approximate circuit model and the exact circuit model with two different induction motors [10] which are described below.

Approximate Circuit Model

In approximate circuit model we use the starting torque, maximum torque and full load torque to determinate the stator resistance, rotor resistance and stator leakage reactance that minimize the error between estimated and manufacturer data. The fitness function and mathematical formulation are computed as follows:

$$F = (f_1)^2 + (f_2)^2 + (f_3)^2 \tag{3.1}$$

where

$$f_1 = \frac{\frac{K_t R_2}{s\left[(R_1 + R_2/s)^2 + X_1^2\right]} - T_{fl}(mf)}{T_{fl}(mf)}$$

$$f_2 = \frac{\frac{K_t R_2}{(R_1 + R_2)^2 + X_1^2} - T_{lr}(mf)}{T_{lr}(mf)}$$

$$f_3 = \frac{\frac{K_t}{2\left[R_1 + \sqrt{R_1^2 + X_1^2}\right]} - T_{max}(mf)}{T_{max}(mf)}$$

$$K_t = \frac{3V_{ph}^2}{\omega_s}$$

Subject to

$$X_{i,min} \leq X_i \leq X_{i,max}$$

where $X_{i,min}$ and $X_{i,max}$ is the lower and upper bound of parameter X_i respectively.

$$\frac{T_{max}(C) - T_{max}(mf)}{T_{max}(mf)} \leq \pm 0.2$$

where $T_{max}(C)$ is the maximum torque calculated (Fig. 3.1).

Fig. 3.1 Approximate circuit
model

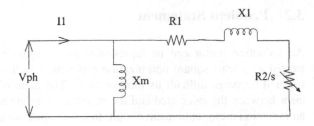

Exact Circuit Model

In the exact circuit model we adjust the stator resistance, rotor resistance, stator leakage resistance stator leakage inductance, rotor leakage reactance and magnetizing leakage reactance to determinate the maximum torque, full load torque, starting torque and full load power factor, the objective function and mathematical formulation are described below,

$$F = (f_1)^2 + (f_2)^2 + (f_3)^2 + (f_4)^2 \tag{3.2}$$

where

$$f_1 = \frac{\frac{K_t R_2}{s\left[(R_{th} + R_2/s)^2 + X^2\right]} - T_{fl}(mf)}{T_{fl}(mf)}$$

$$f_2 = \frac{\frac{K_t R_2}{(R_{th} + R_2)^2 + X^2} - T_{lr}(mf)}{T_{lr}(mf)}$$

$$f_3 = \frac{\frac{K_t}{2\left[R_{th} + \sqrt{R_{th}^2 + X^2}\right]} - T_{max}(mf)}{T_{max}(mf)}$$

$$f_4 = \frac{\cos\left(\tan^{-1}\left(\frac{X}{R_{th} + R_2/s}\right)\right) - pf_{fl}(mf)}{pf_{fl}(mf)}$$

$$V_{th} = \frac{V_{ph} X_m}{X_1 + X_m}, \quad R_{th} = \frac{R_1 X_m}{X_1 + X_m}, \quad X_{th} = \frac{X_1 X_m}{X_1 + X_m}, \quad K_t = \frac{3V_{th}^2}{\omega_s}, \quad X = X_2 + X_{th}$$

Subject to

$$X_{i,min} \leq X_i \leq X_{i,max}$$

$$\frac{T_{max}(C) - T_{max}(mf)}{T_{max}(mf)} \leq \pm 0.2$$

$$\frac{p_{fl} - (I_1^2 R_1 + I_2^2 R_2 + P_{rot})}{p_{fl}} = \eta_{fl}(mf)$$

Fig. 3.2 Exact circuit model

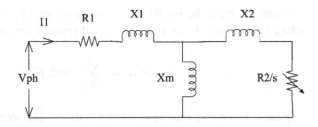

Table 3.1 Manufacturer data

	Motor 1	Motor 2
Capacity (HP)	5	40
Voltage (V)	400	400
Current (A)	8	45
Frequency (Hz)	50	50
No. poles	4	4
Full load slip	0.07	0.09
Starting torque	15	260
Max. torque	42	370
Statin current	22	180
Full load torque	25	190

where P_{rot} is rotational power losses and p_{fl} is the rated power (Fig. 3.2; Table 3.1).

3.3 Gravitational Search Algorithm

GSA was proposed by Rashedi et al. [14] in 2009 based on the law of gravity and mass interactions inspired on Newtonian gravity and the laws of motion. This algorithm uses agents called masses, the masses attract each other with a 'gravitational force' that causes a movement of all masses towards the objects with heavier masses. Now, considering a computational model which has a i-th number of agents defined as follows

$$\mathbf{x}_i = \left(x_i^1, \dots, x_i^d, \dots, x_i^n\right) \quad \text{for } i = 1, 2, \dots, N \qquad (3.3)$$

where x_i^d represents the position of i-th agent in the d-th dimension. At a time t the force acting from a mass i to a mass j is defined as follows

$$F_{ij}^d(t) = G(t)\frac{Mp_i(t) \times Ma_j(t)}{R_{ij}(t) + \varepsilon}\left(x_j^d(t) - x_i^d(t)\right) \qquad (3.4)$$

where Ma_j is the active gravitational mass related to agent j, Mp_i is the passive gravitational of agent i, $G(t)$ is the gravitational constant at time t, ε is a small constant and R_{ij} is the Euclidian distance between the i- and j-th agents. The force acting over an agent i in a d dimensional space is described below

$$F_i^d(t) = \sum_{j=1, j \neq i}^{N} \text{rand}_j \, F_{ij}^d(t) \tag{3.5}$$

Hence, following the Newton's second law the acceleration of the agent i at time t is computed as follows

$$a_i^d(t) = \frac{F_i^d(t)}{Mn_i(t)} \tag{3.6}$$

where Mn_i is the inertial mass of agent i. Therefore the new velocity and position are calculated as follows:

$$\begin{aligned} v_i^d(t+1) &= \text{rand}_i \times v_i^d(t) + a_i^d(t) \\ x_i^d(t+1) &= x_i^d(t) + v_i^d(t+1) \end{aligned} \tag{3.7}$$

The initial value of gravitational constant G, will be changing with time depending to search strategy. Consequently G is a function of the initial value of gravitational constant G_0 and time t:

$$G(t) = G(G_0, t) \tag{3.8}$$

Gravitational and inertia messes are evaluated by a cost function which determinates the quality of the particle, a heavier mass means a better solution. The gravitational and inertia masses are updating by the following equations

$$M_{ai} = M_{pi} = M_{ii} = M_i, \quad i = 1, 2, \ldots, N, \tag{3.9}$$

$$m_i(t) = \frac{fit_i(t) - worst(t)}{best(t) - worst(t)}, \tag{3.10}$$

$$M_i(t) = \frac{m_i(t)}{\sum_{j=1}^{N} m_j(t)}. \tag{3.11}$$

3.4 Experimental Results

In these experiments, it was used the GSA to determinate the optimal parameters of two induction motors considering the approximate circuit model and exact circuit model. We also use DE, ABC and PSO to solve the same application, in order to compare and validate the results obtained by GSA, due these algorithms are widely used in literature showing good performance. The parameter used for each algorithm in this experiment are mentioned below

1. PSO, parameters $c_1 = 2$, $c_2 = 2$ and weights factors were set $w_{max} = 0.9$, and $w_{min} = 0.4$ [23].
2. ABC, the parameters implemented were provided by [24], *limit* = 100.
3. DE, in accordance with [25] the parameters were set $p_c = 0.5$ and $f = 0.5$.
4. GSA, the parameter were set according to [14].

3.4.1 Induction Motor Parameter Identification

For each algorithm, it was considered a population size of 25 and 3000 iterations. To carry out the experiment 35 independent trials were performed. Fitness value, deviation standard and mean of each algorithm for the approximate circuit model and motor 1 is reported in Table 3.2, using the approximate circuit model with the motor 2 is shown in Table 3.3, for the exact model and motor 1 is given in Table 3.4 and for the exact model and motor 2 the results are reported in Table 3.5.

Table 3.2 Fitness value of approximate circuit model, motor 1

	GSA	DE	ABC	PSO
Min	3.4768e−22	1.9687e−15	2.5701e−05	1.07474e−04
Max	1.6715e−20	0.0043	0.0126	0.0253
Mean	5.4439e−21	1.5408e−04	0.0030	0.0075
Std	4.1473e−21	7.3369e−04	0.0024	0.0075

Table 3.3 Fitness value of Approximate circuit model, motor 2

	GSA	DE	ABC	PSO
Min	3.7189e−20	1.1369e−13	3.6127e−04	0.0016
Max	1.4020e−18	0.0067	0.0251	0.0829
Mean	5.3373e−19	4.5700e−04	0.0078	0.0161
Std	3.8914e−19	0.0013	0.0055	0.0165

Table 3.4 Fitness value of exact circuit model, motor 1

	GSA	DE	ABC	PSO
Min	0.0032	0.0172	0.0172	0.0174
Max	0.0032	0.0288	0.0477	0.0629
Mean	0.0032	0.0192	0.0231	0.0330
Std	0.0000	0.0035	0.0103	0.0629

Table 3.5 Fitness value of exact circuit model, motor 2

	GSA	DE	ABC	PSO
Min	**0.0071**	0.0091	0.0180	0.0072
Max	**0.0209**	0.0305	0.2720	0.6721
Mean	**0.0094**	0.0190	0.0791	0.0369
Std	**0.0043**	0.0057	0.0572	0.1108

Table 3.6 Comparison of GSA, DE, ABC and PSO with manufacturer data, approximated circuit model, motor 1

	True-val	GSA	Error (%)	DE	Error (%)	ABC	Error (%)	PSO	Error (%)
Tst	15	**15.00**	**0**	14.9803	−0.131	14.3800	−4.133	15.4496	2.9973
Tmax	42	**42.00**	**0**	42.0568	0.135	40.5726	−3.398	39.6603	−5.570
Tfl	25	**25.00**	**0**	24.9608	−0.156	25.0480	0.192	25.7955	3.182

Table 3.7 Comparison of GSA, DE, ABC and PSO with manufacturer data, approximated circuit model, motor 2

	True-val	GSA	Error (%)	DE	Error (%)	ABC	Error (%)	PSO	Error (%)
Tst	260	**260.00**	**0**	258.4709	−0.588	260.6362	0.2446	288.9052	11.117
Tmax	370	**370.00**	**0**	372.7692	0.7484	375.0662	1.3692	343.5384	−7.151
Tfl	190	**190.00**	**0**	189.0508	−0.499	204.1499	7.447	196.1172	3.2195

After evaluating the parameters determined by each algorithm the results were compared with manufacturer data taken from the Table 3.1. It is reported the approximate model with model 1 and motor 2, exact model with motor 1 and motor 2 in Tables 3.6, 3.7, 3.8 and 3.9 respectively. The convergence diagram is plotted in

Table 3.8 Comparison of GSA, DE, ABC and PSO with manufacturer data, exact circuit model, motor 1

	True-val	GSA	Error (%)	DE	Error (%)	ABC	Error (%)	PSO	Error (%)
Tst	15	**14.9470**	**−0.353**	15.4089	2.726	16.4193	9.462	15.6462	4.308
Tmax	42	**42.00**	**0**	42.00	0	42.00	0	42.00	0
Tfl	25	**25.0660**	**0.264**	26.0829	4.3316	25.3395	1.358	26.6197	6.4788

Table 3.9 Comparison of GSA, DE, ABC and PSO with manufacturer data, exact circuit model, motor 2

	True-val	GSA	Error (%)	DE	Error (%)	ABC	Error (%)	PSO	Error (%)
Tst	260	**258.1583**	**−0.708**	262.0565	0.7909	246.2137	−5.302	281.8977	8.4221
Tmax	370	**370.00**	**0**	370.00	0	370.00	0	370.00	0
Tfl	190	**189.8841**	**−0.061**	192.2916	1.2061	207.9139	9.428	166.6764	−12.27

Fig. 3.3 which shows the evolution of each algorithm through iterations in exact circuit model with the motor 2. And finally, the curve generated by the slip versus torque in both models (Figs. 3.4 and 3.5) with motor 1 and motor 2 are shown in Figs. 3.6 and 3.7 respectively. From Tables 3.2 to 3.5 are depicted in bold case the minimum values resulting from the optimization procedure. In the case of Tables 3.6 to 3.9 are depicted in bold case the minimum values resulting from the optimization procedure, the minimum error values and the true-value (original value to compare), in the cases where two or more algorithms obtained the same value, there are all depicted in bold letters.

$$
\boxed{
\begin{array}{l}
\text{Randomized Initialization of population} \\
\text{Find the best solution in the initial population} \\
\textbf{wile } (\textit{stop criteria}) \\
\quad \textbf{for } i{=}1{:}N \text{ (for all agents)} \\
\qquad \text{update } G(t),\ best(t),\ worst(t) \text{ and } M_i(t) \text{ for } i = 1, 2.., N \\
\qquad \text{calculate the mass of each agent } M_i(t) \\
\qquad \text{calculate the gravitational constant } G(t) \\
\qquad \text{calculate acceleration in the gravitational field } a_i^{d}(t) \\
\qquad \text{update the velocity and positions of the agents } v_i^{d},\ x_i^{d} \\
\quad \textbf{end } (\text{for}) \\
\qquad \text{Find the best solution} \\
\textbf{end } (\text{while}) \\
\qquad \text{Display the best solution}
\end{array}
}
$$

Fig. 3.3 Gravitational search algorithm (GSA) pseudo code

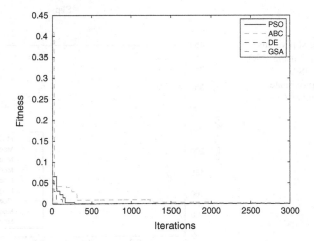

Fig. 3.4 Convergence evolution through iterations of model 1

Fig. 3.5 Convergence
evolution through iterations of
model 2

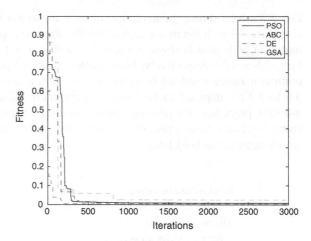

Fig. 3.6 Curve slip versus
torque of motor 1 using PSO,
ABC, DE and GSA
considering approximate
model circuit and exact model
circuit

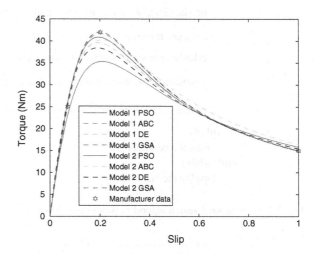

Fig. 3.7 Curve slip versus
torque of motor 2 using PSO,
ABC, DE and GSA
considering approximate
model circuit and exact model
circuit

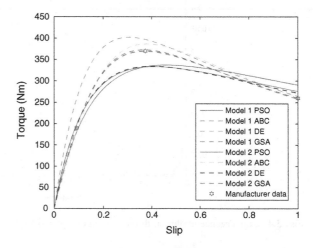

3.4.2 Statistical Analysis

After 35 independents executions of each evolutionary technique, we proceed to validate the results using a non-parametric statistical analysis know as Wilcoxon's rank sum [26] which consider a 0.05 of significance between the "average $f(\theta)$ value" to determinate if there is difference. Table 3.10 shows the values of P in the comparison of the presented technique, GSA versus DE, GSA versus ABC and GSA versus PSO where the null hypothesis is the value of significance higher to 5% that indicates there is no difference enough between samples.

For all cases it is possible to perceive, that the significance value is lower than 0.05, this indicates that the null hypothesis was rejected demonstrating that exist difference enough between results given by GSA and the algorithms used in this work for the comparison, such evidence suggest that GSA surpasses to the most common optimization techniques for the consistency of its results through an efficient search strategy and has not occurred by coincidence but the algorithm is able to find efficient solution due its robustness and accuracy.

3.5 Conclusions

This chapter presents the GSA to determinate induction motor parameters of approximate circuit model and exact circuit model using two different induction motors. The estimation process of an induction motor parameters is treated as a least square optimization problem which becomes in a complex task for the non-linearity of the steady-state equivalent circuit model.

The presented scheme outperforms the most popular optimization algorithms such as DE, ABC and PSO, minimizing the error between the calculated and manufacturer data and converging faster than the other techniques used for the comparison. After 35 individual executions of each algorithm, we used a non-parametric statistical validation known as Wilcoxon's rank sum, which proved

Table 3.10 P-value from Wilcoxon's rank sum test of the comparison of GSA, DE, ABC and PSO

Model/motor	GSA versus		
	DE	ABC	PSO
Model 1, motor 1	6.545500588914223e−13	6.545500588914223e−13	6.545500588914223e−13
Model 1, motor 2	0.009117078811112	0.036545600995029	0.004643055264741
Model 2, motor	6.545500588914223e−13	6.545500588914223e−13	6.545500588914223e−13
Model 2, motor 2	1.612798082388261e−09	9.465531545379272e−13	3.483016312301559e−08

that there is indeed a significant difference between the results obtained by GSA and the techniques used for the comparison, improving the results reported in literature and showing good performance in complex applications and consistency of its solutions by operators used in the search strategy.

References

1. Çaliş H, Çakir A, Dandil E (2013) Artificial immunity-based induction motor bearing fault diagnosis. Turk J Electr Eng Comput Sci 21(1):1–25
2. Prakash V, Baskar S, Sivakumar S, Krishna KS (2008) A novel efficiency improvement measure in three-phase induction motors, its conservation potential and economic analysis. Energy Sustain Dev 12(2):78–87
3. Waters SS, Willoughby RD (1983) Modeling induction motors for system studies. IEEE Trans Ind Appl IA-19(5):875–878
4. Ansuj S, Shokooh F, Schinzinger R (1989) Parameter estimation for induction machines based on sensitivity analysis. IEEE Trans Ind Appl 25(6):1035–1040
5. De Kock J, Van der Merwe F, Vermeulen H (1994) Induction motor parameter estimation through an output error technique. IEEE Trans Energy Convers 9(1):69–76
6. Mohammadi HR, Akhavan A (2014) Parameter estimation of three-phase induction motor using hybrid of genetic algorithm and particle swarm optimization, J Eng, vol 2014
7. Bishop RR, Richards GG (1990) Identifying induction machine parameters using a genetic optimization algorithm. In: Proceedings on IEEE South east conference, vol 2, pp 476–479
8. Lindenmeyer D, Dommel HW, Moshref A, Kundur P (2001) An induction motor parameter estimation method. Int J Electr Power Energy Syst 23(4):251–262
9. Sakthivel VP, Bhuvaneswari R, Subramanian S (2010) An improved particle swarm optimization for induction motor parameter determination. Int J Comput Appl 1(2):71–76
10. Sakthivel VP, Bhuvaneswari R, Subramanian S (2010) Artificial immune system for parameter estimation of induction motor. Expert Syst Appl 37(8):6109–6115
11. Sakthivel VP, Bhuvaneswari R, Subramanian S (2011) An accurate and economical approach for induction motor field efficiency estimation using bacterial foraging algorithm. Meas J Int Meas Confed 44(4):674–684
12. Perez I, Gomez-Gonzalez M, Jurado F (2013) Estimation of induction motor parameters using shuffled frog-leaping algorithm. Electr Eng 95(3):267–275
13. Abro AG, Mohamad-Saleh J (2014) Multiple-global-best guided artificial bee colony algorithm for induction motor parameter estimation. Turk J Electr Eng Comput Sci 22:620–636
14. Rashedi E, Nezamabadi-pour H, Saryazdi S (2009) GSA: a gravitational search algorithm. Inf Sci (NY) 179(13):2232–2248
15. Farivar F, Shoorehdeli MA (2016) Stability analysis of particle dynamics in gravitational search optimization algorithm. Inf Sci 337:25–43
16. Yazdani S, Nezamabadi-pour H, Kamyab S (2014) A gravitational search algorithm for multimodal optimization. Swarm Evol Comput 14:1–14
17. Beigvand SD, Abdi H, La Scala M (2016) Combined heat and power economic dispatch problem using gravitational search algorithm. Electr Power Syst Res 133:160–172
18. Kumar V, Chhabra JK, Kumar D (2014) Automatic cluster evolution using gravitational search algorithm and its application on image segmentation. Eng Appl Artif Intell 29:93–103
19. Zhang W, Niu P, Li G, Li P (2013) Forecasting of turbine heat rate with online least squares support vector machine based on gravitational search algorithm. Knowl Based Syst 39:34–44
20. Karaboga D (2005) An idea based on honey bee swarm for numerical optimization. Technical Report TR06, Erciyes University Press, Erciyes, p 10

21. Storn R, Price K (1997) Differential evolution—a simple and efficient heuristic for global optimization over continuous spaces. J Glob Optim 11:341–359
22. Kennedy J, Eberhart R (1995) Particle swarm optimization. In: Proceedings of IEEE international conference on neural networks, Piscataway, vol 4, pp 1942–1948
23. Sakthivel VP, Subramanian S (2011) On-site efficiency evaluation of three-phase induction motor based on particle swarm optimization. Energy 36(3):1713–1720
24. Jamadi M, Merrikh-Bayat F (2014) New method for accurate parameter estimation of induction motors based on artificial bee colony algorithm. arXiv:1402.4423
25. Ursem RK, Vadstrup P (2003) Parameter identification of induction motors using differential evolution. In: The 2003 congress on evolutionary computation, CEC'03, vol 2, pp 790–796
26. Kotz S, Johnson NL (eds) (1992) Breakthroughs in statistics: methodology and distribution. Springer, New York, NY, USA, pp 196–202

Chapter 4
Color Segmentation Using LVQ Neural Networks

Color segmentation in digital images is a challenging task due to image capture conditions. Typical segmentation algorithms present several difficulties in this process because they do not tolerate variations in color hue corresponding to the same object. In this chapter is presented an approach for color segmentation based on learning vector quantization (LVQ) networks, which conducts the segmentation process by means a color-based pixel classification. In the LVQ networks, neighboring neurons have the capability to learn how to recognize close sections of the input space. The presented segmentation approach classifies the pixels directly by means of the LVQ network. The experimental results over a set of images show the efficiency of the LVQ-based method to satisfactorily segment color despite remarkable illumination problems.

4.1 Introduction

The color discrimination plays an important role in humans for individual object identification. Humans usually do not search in a bookcase for a previously known book solely by its title. We try to remember the color on the cover (e.g., blue) and then search among all the books with a blue cover for the one with the correct title. The same applies to recognizing an automobile in a parking site. In general, humans do not search for model A of company B, but rather we look for a red car. It is only when a red vehicle is spotted, when it is decided according to its geometry, whether that vehicle is the one of the required kind.

Image segmentation is the first step in image analysis and pattern recognition. It is a critical and essential component but also it is one of the most difficult tasks in image processing. The actual operation of the algorithm determines the quality of the overall image analysis.

Color image segmentation is a process of extracting from the image domain one or more connected regions satisfying the uniformity (homogeneity) criterion [1]

© Springer International Publishing AG 2017   59
M.-A. Díaz-Cortés et al., *Engineering Applications of Soft Computing*,
Intelligent Systems Reference Library 129, DOI 10.1007/978-3-319-57813-2_4

which is derived from spectral components [2, 3]. These components are defined within a given color space model such as the RGB model-the most common model, which considers that a color point is defined by the color component levels of the corresponding pixel, i.e. red (R), green (G), and blue (B). Other color spaces can also be employed considering that the performance of an image segmentation procedure is known to depend on the choice of the color space. Many authors have sought to determine the best color space for their specific color image segmentation problems. Unfortunately, there is not an ideal color space to provide satisfying results for the segmentation of all kinds of images.

Image segmentation has been the subject of considerable research activity over the last two decades. Many algorithms have been elaborated for gray scale images. However, the problem of segmentation for color images that implies a lot of information about objects in scenes has received much less attention of the scientific community. Although color information allows a more complete representation of images and more reliable segmentations, processing color images requires computational times considerably larger than those needed for gray-level images as it is very sensitive to illumination changes.

This chapter considers the color image segmentation as a pixel classification problem. By means of the LVQ neural networks and their classification schemes, classes of pixels are detected by analyzing the similarities between the colors of the pixels.

In particular, color image segmentation techniques described in the literature can be categorized into four main approaches: Histogram thresholding and color space clustering; region based approaches, edge detection, probabilistic methods and soft-computing techniques. The following section discusses on each techniques, summarizing their main features.

4.1.1 Histogram Thresholding and Color Space Clustering

Histogram thresholding is one of the widely used techniques for monochrome image segmentation. It assumes that images are composed of regions with different gray levels. The histogram of an image can be separated into a number of peaks (modes), each corresponding to one region, and there exists a threshold value corresponding to valley between the two adjacent peaks. As for color images, the situation is different from monochrome images because of multi-features. Multiple histogram-based thresholding divides the color space by thresholding each component histogram.

The classes for color segmentation are built by means of a cluster identification scheme which is performed either by an analysis of the color histogram [4] or by a cluster analysis procedure [5]. When the classes are constructed, the pixels are assigned to one of them by means of a decision rule and then mapped back to the original image plane to produce the segmentation. The regions of the segmented image are composed of connected pixels which are assigned to the same classes.

When the distribution of color points is analyzed in the color space, the procedures generally lead to a noisy segmentation with small regions scattered through the image. Usually, a spatial-based post-processing is performed to reconstruct the actual regions in the image [6].

4.1.2 Edge Detection

Region based approaches, including region growing, region splitting [7], region merging [8] and their combination [9], attempt to group pixels into homogeneous regions. In the region growing approach, a seed region is first selected. Thus, it is expanded to include all homogeneous neighbors, repeating the process until all pixels in the image are classified. One problem with region growing is its inherent dependence on the selection of seed region and the order in which pixels and regions are examined. In the region splitting approach, the initial seed region is simply the whole image. If the seed region is not homogeneous, it is usually divided into four squared sub-regions, which become new seed regions. This process is repeated until all sub-regions are homogeneous. The major drawback of region splitting is that the resulting image tends to mimic the data structure used to represent the image and comes out too square. The region merging approach is often combined with region growing or region splitting to merge similar regions for making a homogeneous section as large as possible.

4.1.3 Probabilistic Methods

Probabilistic color segmentation estimates the probability $P_i(x, y) \in [0, 1]$ for a given pixel $I(x, y)$ of belonging to a region i in the image I. Although the probability density $P_i(x, y)$ is usually determined, its parameters are often unknown. In [10–12] have already discussed color segmentation when the joint distribution of color is modeled by a mixture of Gaussians within a 3-dimensional space. Since no spatial coordinates are incorporated, once the model has been inferred, it needs a spatial grouping step which applies a maximum-vote filter and uses the connected component algorithm.

Isard and MacCormick [13] have employed color information to implement particle filtering. Lately, Perez et al. [14] introduced an approach that also uses color histograms and particle filtering for multiple object tracking. Both methods differ in the initialization procedure for the tracker, the model updating, the region shape and the observation of the tracking performance. Bradski [15] modified the mean-shift algorithm (Camshift) which operates on probability distributions to track colored objects in video frame sequences.

4.1.4   Soft-Computing Techniques

A trendy issue is the use of soft-computing approaches for image processing systems. Artificial neural network models have been proposed to segment images directly from pixel similarity or discontinuity. More than 200 neural networks used in image processing are presented by [1] by means of an 2D taxonomy. In [2] it is also discussed on many color image segmentation techniques, including the histogram thresholding, characteristic feature clustering, edge detection, region-based methods, fuzzy methods, and neural networks.

Color segmentation is successfully computed by self-organizing maps (SOMs) and competitive networks in [16–18]. In [17] a two-stage strategy includes a fixed-size two dimensional feature map (SOM) to capture the dominant colors of an image by unsupervised training. In a second stage, the algorithm combines a variable-sized one-dimensional feature map and color merging to control the number of color clusters that are used for segmentation. The model in [16] is based on a two-step neural network. In the first step, a SOM performs color reduction and then a simulated annealing step searches for the optimal clusters from SOM prototypes. The task involves a procedure of hierarchical prototype learning (HPL) to generate different sizes of color prototypes from the sampled object colors.

4.1.5   Scheme

Learning vector quantization (LVQ) networks learn to recognize groups of similar input vectors in such a way neurons that locate nearby to others in the neuron layer respond to similar input vectors. The learning is supervised and the inputs vectors into target classes are chosen by the user.

The LVQ algorithm presented in this chapter works only with image pixels, with no dynamic model or probability distribution, which in turn, improves the processing speed and facilitates the implementation process. The approach naturally avoids the complex structures commonly resulting from other neural methods such as those in [16–18]. It incorporates a decision function which eases the segmentation of the objective color. The method has been applied on several color segmentation problems (face localization and color tracking), showing enough capacity to comprehensively segment color even under illumination differences.

This chapter is organized as follows: Sect. 4.2 revisits some background concepts while Sect. 4.3 presents an introductory study of competitive neural networks and their main features. Section 4.4 explains relevant details of LVQ networks and Sect. 4.5 shows the architecture and characteristics of the presented color-segmentation system, including some practical discussions. Section 4.6 offers a simple explanation on the algorithm's implementation. Section 4.7 reports the results and some discussions, finally in Sect. 4.8 are presented some conclusions.

4.2 Background Issues

4.2.1 RGB Space Color

Color is perceived by humans as a combination of triple stimuli R (red), G (green), and B (blue) which are usually named as primary colors. From R, G, B representation, it is possible to derive other kinds of color representations (spaces) by using either linear or nonlinear transformations. The RGB color space can be geometrically represented within a 3-dimensional cube as shown in Fig. 4.1. The coordinates of each point inside the cube represent the values of red, green and blue components, respectively.

The laws of color theory are: (1) any color can be created by these three colors and the combination of the three colors is unique; (2) if two colors are equivalent, they will be again equivalent after multiplying or dividing the three components by the same number; (3) the luminance of a mixture of colors is equal to the sum of the luminance of each color. The triple stimuli values that served as the color basis are: 425.8 nm for blue, 546.1 nm for green and 700.0 nm for red. Any color can be expressed by these three color bases.

RGB is the most widely accepted model for television systems and pictures acquired by digital cameras. Video monitors commonly display color images by modulating the intensity of the three primary colors (red, green, and blue) at each pixel of the image [19]. RGB is suitable for color display as it is complicated for color segmentation's purposes, considering the high correlation among the R, G, and B components [20]. High correlation refers to the intensity changes which assume that all the three components will change accordingly. The measurement of a color in RGB space does not represent color differences in a uniform scale and hence it is impossible to evaluate the similarity of two colors from their distance in RGB space.

4.2.2 Artificial Neural Networks

Artificial Neural Networks are composed from simple elements that commonly mimic biological systems following parallel arrangements. By nature, a network

Fig. 4.1 RGB space color

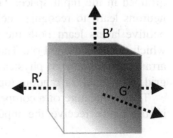

Fig. 4.2 Supervised learning
in neural networks

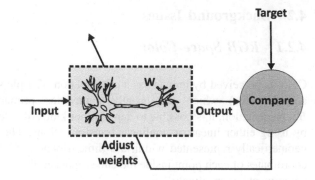

function is determined by the connections between such neural elements. It is possible to train a neural network to "learn" a given function by adjusting the values of the connections (**W** weights) between elements.

A common training algorithm seeks to match a given neural input to a specific target output as shown in Fig. 4.2. The network is adjusted by comparing the network's output and the target value, until the network output matches, as close as possible, the target. Typically, a great number of input/target pairs are used following the *supervised learning* scheme to train the network.

Batch training of a network proceeds by making weight and bias changes based on an entire set (batch) of input vectors. Incremental training changes are applied to the weights and biases of a network after the presentation of each individual input vector. Incremental training is sometimes referred as on-line or adaptive training.

Neural networks may be employed to solve several sorts of problems, ranging from pattern recognition, identification, classification, speech, control systems and computational vision. The supervised training methods are widely known in the school. Other kind of networks can be obtained from *unsupervised training* techniques or from direct design methods. Unsupervised networks can be applied, for instance, to identify groups of data.

4.3 Competitive Networks

Competitive Networks [21] learn to classify input vectors according to how they are grouped in the input space. They differ from other networks in that neighboring neurons learn to recognize neighboring sections of the input space. Thus, competitive layers learn both the distributions and topology of the input vectors in which they are trained on. The neurons in the layer of a competitive network are arranged originally in physical positions according to a topology pattern such as grid, hexagonal, or random topology.

The architecture of a competitive network is shown in Fig. 4.3. The |*Ndist*| box in the figure receive the input vector **p** and the input weight matrix **IW** and

Fig. 4.3 Architecture of a competitive network

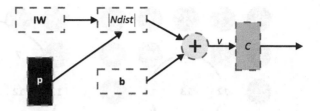

produces a vector or a matrix **S** according to the topological configuration. The elements are the negative distances between the input vector **p** and the matrix **IW**. The net value v of the competitive layer is computed by finding the negative distance between input vector **p** and the weight matrix **IW** and then adding the biases **b**. If, all biases are zero, the maximum net input that a neuron can have is 0. This occurs when the input vector **p** equals the neuron's weight vector contained in the matrix **IW**.

The competitive transfer function C receive a net value v and returns outputs of 0 for all neurons except for the *winner*, the neuron associated with the most positive element of input v. Thus, the winner's output is 1. The weights of the winning neuron are adjusted by the Kohonen learning rule. Supposing that the ith neuron wins, the elements of the ith row of the input weight matrix and all neurons within a certain neighborhood radius $Ni(d)$ of the winning neuron are adjusted as shown in Eq. (4.1). In other words, $Ni(d)$ is the neighbor's number around of the winner neuron to be affected.

$$_i\mathbf{IW}^{1,1}(q) = {}_i\mathbf{IW}^{1,1}(q-1) + \alpha(\mathbf{p}(q) - {}_i\mathbf{IW}^{1,1}(q-1)) \tag{4.1}$$

Here α is the learning rate and $Ni(d)$ contains the index for all of the neurons that lie within a radius d of the ith winning neuron. Thus, when a vector **p** is presented, the weights of the winning neuron and its closest neighbors move toward **p**. Consequently, after many presentations, neighboring neurons will have learned vectors similar to each others. The winning neuron's weights are altered accordingly by the learning rate. The weights of neurons in its neighborhood are altered proportional to half of the learning rate. In this work, the learning rate and the neighborhood distance (used to determine which neurons are in the winning neuron's neighborhood) are not altered during training.

To illustrate the concept of neighborhoods, consider the Fig. 4.4. Left, it is shown a two dimensional neighborhood of radius $d = 1$ around neuron 13. Aside it is shown a neighborhood of radius $d = 2$.

These neighborhoods could be written as:

$$N_{13}(1) = (8, 12, 13, 14, 18), N_{13}(2) = (3, 7, 8, 9, 11, 12, 13, 14, 15, 17, 18, 19, 23)$$

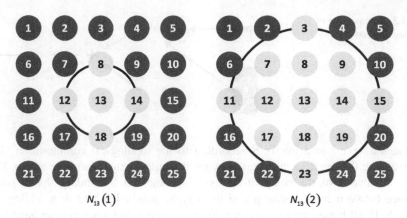

Fig. 4.4 *Left*, two dimensional neighborhood with radius $d = 1$. *Right*, neighborhood with radius $d = 2$

4.4 Learning Vectors Quantization Vectors

An LVQ network [22] has first, a competitive layer and second, a linear layer. The competitive layer learns to classify input vectors like the networks of the last section. The linear layer transforms the competitive layer's classes into target classifications defined by the user. We refer to the classes learned by the competitive layer as *subclasses* and the classes of the linear layer as *target classes*. Both the competitive and linear layers have one neuron per class. However, the neurons in the competitive layer can be arranged according to a topology pattern.

Thus, the competitive layer can learn **S1** classes, according to how they are grouped in the topological space. These, in turn, are combined by the linear layer to form **S2** target classes. This process can be considered as a lineal transformation carried out on the learned classes **S1** (in unsupervised manner) by the competitive layer to a mapping on **S2** defined by **LW**. This transformation allows distributing similar patterns around the target neuron in the linear layer. The LVQ network architecture is shown in Fig. 4.5.

4.5 Architecture of the Color Segmentation System

The core of the proposed algorithm is a LVQ network whose inputs are connected directly to each RGB pixel component of the image **I**. The output of the LVQ network is a vector **S2** connected to the decision function $\mathbf{f_d}$. If the RGB components of the original pixel represents the color to be segmented, then the $\mathbf{f_d}$ function output is 1, if not is 0. The result is a new image **I′**. The segmentator takes

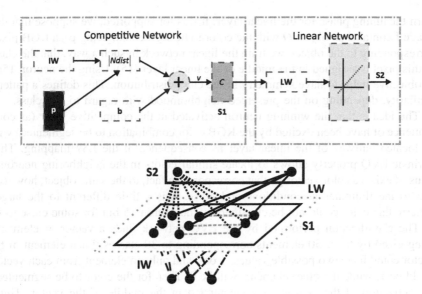

Fig. 4.5 Schematic representations of the LVQ net

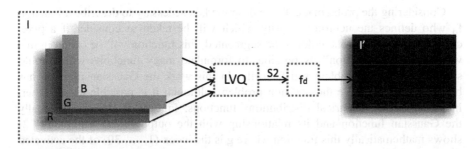

Fig. 4.6 Architecture of the color segmentation system

advantage of the LVQ property to learn to recognize neighboring sections of the input space. Figure 4.6 shows the segmentator's architecture.

Considering that the LVQ net is configured with a grid of 6×5 neurons in the competitive layer and 30 one-dimensional output neurons (linear layer), then would be possible to train the competitive network to learn the color-pixel space and its topology (described as the vector $\mathbf{p}$ with elements $\mathbf{p}_R$, $\mathbf{p}_G$ and $\mathbf{p}_B$ coming from the image). The 6×5 grid in the competitive layer was chosen after considering a tradeoff between the neuron distribution and the computational cost [17]. The size of the linear layer (30) is considered only as being coherent to the neurons contained on the grid (6×5).

The net training is achieved in two phases: Ordering phase and tuning phase. In the ordering phase the neurons of competitive layer learn to recognize groups of similar color vectors in an *unsupervised* manner. Using *supervised* learning, they learn the tuning phase for the linear layer. It was for supported, we suppose that the image **I** contains an object O with the color to be segmented, being $\mathbf{p}_O$ a RGB pixel corresponding to the object, we train the linear network in such a way, that the class of this pixel is assigned in the middle of the linear layer (15). Using the neuron 15th as objective helps to have symmetry in the class distribution. This defines a pattern similarity, depending on the presented neighborhood with regard to this class.

The idea is that the winning neuron activated in the competitive layer (as consequence of have been excited by the RGB color combination to be segmented) will be located halfway of the linear layer as consequence of the **LW** mapping. This intrinsic LVQ property allows to locate similar colors in the neighboring neurons. Thus, if exists a color vector $\mathbf{p}_1$ that correspond, in fact, to the same object, however due to the illumination conditions and noise, it is a little different to the target pattern, this color could not be classified by the neuron 15 but for some close to it.

The classification performed by the LVQ network finds a vector of elements categorized by $\mathbf{S}^2$ of 30 elements corresponding to 30 classes. Each element of $\mathbf{S}^2$ vector could have two possible values, 1 or 0 and only an element from each vector could be 1, while the other elements will be 0. Then for the color to be segmented, the activation of the neurons is concentrated in the middle of the vector, Thus, neurons nearest to the 15 will have a bigger possibility to be activated, for similar color patterns.

Considering the problematic above described, is necessary to describe a function $\mathbf{f_d}$ who defines the neuron's density which will be taken to consider if a pixel corresponds or not to the color to be segmented, this function will be called in this work "decision function". Is possible to formulate many functions which could solve the decision problem satisfactorily. In this work the Gaussian function has been chosen to resolve the decision problem, although it is possible to use other, including non-symmetrical distributions functions. Figure 4.7 shows graphically the Gaussian function and its relationship with the output layer. Equation (4.2) shows mathematically this function where $\mathbf{g}$ is the index (1, …, 30) of the activated neuron, N is the neuron number of the linear layer (for this paper, N = 30) and σ is the standard deviation. Therefore, $\mathbf{f_d}$ has only a calibration parameter represented by σ which determines the generalization capacity of the complete system. Thus, for example, if the value of σ is chosen small enough, the segmentation capacity will be more selective than in the case of a bigger σ.

$$\mathbf{f_d(g)} = \frac{1}{\sqrt{2\pi}\sigma}\exp\left(-\frac{\left(\mathbf{g}-(N/2)^2\right)}{2\sigma^2}\right) \tag{4.2}$$

4.6 Implementation

The implementation is divided in two parts, the net training and the segmentator application.

Fig. 4.7 Decision function model

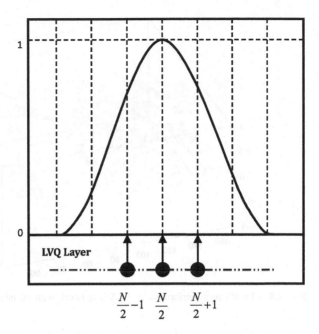

First, the training process requires an image frame containing the object whose color will be segmented. Then, a pixels block is selected to train the LVQ net according to the color to be segmented but specifying that this pixel must be located at the 15th neuron that means, at the middle of the 30 neurons array. Using this selection, the patterns are similarly distributed around the 15th neuron.

The weight matrix **IW** and **LW** are randomly initialized for the training. During the unsupervised training the learning rate $\alpha = 0.1$ was used as well as a distance radius $Ni(d) = 3$. This conditions assure that winner neurons weights will be affected in a reason of 0.1. While its three neighboring neurons weights (in both ways) be affected in a reason of 0.05. During the training of the lineal layer the values of **LW** are calculated. These values transform the classes lineally allowing distributing them around the neuron 15. Initially due to the aleatory configuration, the learned classes by the competitive layer cannot be correctly mapped to the output vector, however after some iteration; these values are classified correctly as consequence of the adaptation of **LW**. Figure 4.8 shows the classification results of the neurons in the competitive layer arranged in a 5×2 grid on an image obtained of Webcam.

For the segmentator use, a decision function $\mathbf{f_d}$ with parameters $\mu = 15$ and $\sigma = 3$ was integrated to the previously trained network. The complete system was programmed using Visual C++ and tested on a PC×86 at 900 MHz with 128 MBytes RAM. Figure 4.9 shows an image with irregular illumination and different images obtained during the segmentation process of the flesh color. The resulting images correspond to the classes represented by the 15th neuron (Fig. 4.9b) and the

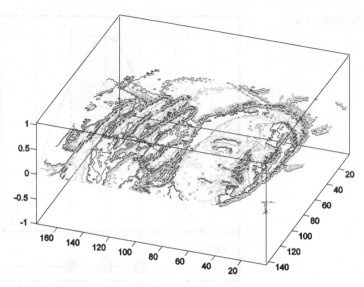

Fig. 4.8 Classification performed by a LVQ network with 10 neurons in the competitive layer

14th neighbor neuron (Fig. 4.9c), as well as the output of the decision function (Fig. 4.9d).

4.7   Results and Discussion

In order to test the robustness of the segmentation algorithm, the approach is applied to video-streamed images. The patch of color to be segmented may exhibit illumination changes within the video sequence; however, the algorithm's operation prevails despite such changes.

The overall performance is tested considering two different experiments. First, the system performance is tested over an indoor footage as shown by Fig. 4.10. In order to train the LVQ Network, one video-frame is considered to choose some pixels that belong to the skin of a couple of human individuals who are participating in a handshaking gesture. Once the system is fully trained, a full-long video sequence is tested. It is important to consider all changes in the skin color that are related to variable illumination. The effect is generated as a result of changes in the relative position and orientation of the object with respect to the light source. Despite such changes and the closeness of other similar colors in the scene, the algorithms have been able to acceptably segment the skin color as shown by Fig. 4.10.

The second experiment tests the algorithm's sensitivity in response to abrupt changes on illumination. An external setting is used to generate the video sequence which naturally includes changes in sun lighting and therefore shading too.

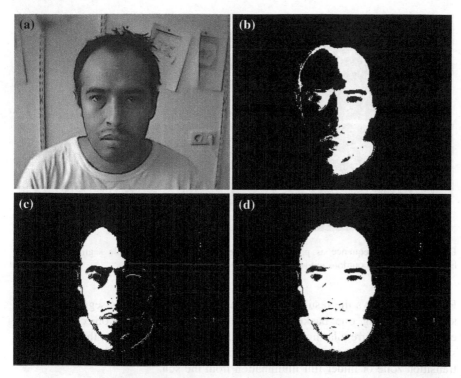

Fig. 4.9 Result Images. **a** Original image, **b** image of the pixels corresponding to the class represented by the 15th neuron, **c** image of the pixels corresponding to the class represented by the 14th neuron and **d** output of the decision function

Fig. 4.10 Sequence performed by the LVQ segmentator in indoor environment

Fig. 4.11 Image sequence as processed by the LVQ segmentator while segmenting for the backpack within an outdoor environment

Figure 4.11 shows the resulting sequence once the algorithm is segmenting a red spot representing one backpack. Although the shape of the object of interest exhibits several irregular holes generated from lighting variations, the algorithm is still capable to segment the patch yet if it is located under the shade, or passing by a transition zone or under full illumination from the sun.

The overall performance of the algorithm can be successfully compared to other algorithms such as those in [23, 24]. However, the LVQ algorithm works directly on the image pixels with no dynamical model or probability distribution, improving the execution time and simplifying the implementation. However, contrary to other approaches [25], the average execution time required by the algorithm to classify one pixel is always constant depending solely on the color complexity.

It is remarkable the influence of the σ parameter over the decision function itself. Best results were achieved with σ falling between 3 and 5. It is easy to envision that improving the robustness of the system may evolve considering an adaptable σ parameter. As an example, Fig. 4.12a, b show several cases when using different values for σ.

The proposed algorithm also exhibits better generalization properties compared to other classical lookup table algorithms. In particular, if it considers images with changing illumination. As shown by Fig. 4.9, the LVQ algorithm is capable of segmenting one face completely while one lookup-table-based algorithm, in the best case, merely reaches similar images to those generated by neighboring output neurons (Fig. 4.9b or c of the LVQ net).

The robustness of the LVQ algorithm facing variable lighting can be compared to the popular CamShift algorithm. Aiming for a fair comparison, three indexes are considered just as it is proposed in [26]. The first performance index is related to "tracking failure". Both algorithms run until a failure in the tracking process is registered. According to this measurement, a track is assumed to be lost either when

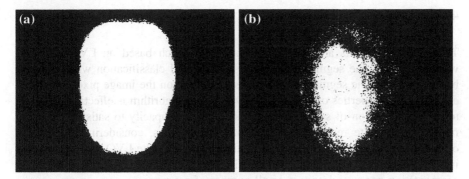

Fig. 4.12 Results obtained using **a** $\sigma = 4$ and **b** $\sigma = 1$

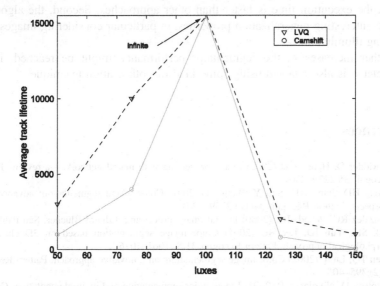

Fig. 4.13 Comparison of LVQ and Camshift algorithms

the center of the segmented object is sufficiently far from the true center, i.e. when the object position determined by the algorithm does not match to the real position or when the computed center falls outside the image location in five consecutive steps. Finally, the "track-lifetime" is defined according to the number of time steps until the tracking was lost. This index is commonly averaged across all trials. Figure 4.13 shows the results from these experiments. Both algorithms were tested considering 100 lx as an initial point—a standard office's illumination. At this point, the averaged track-lifetime is infinite for both algorithms. Measurements are taken in both directions, showing that the LVQ algorithm has a higher robustness, particularly in case of low intensity lighting.

4.8   Conclusions

This chapter presents an image segmentator approach based on LVQ networks which considers the segmentation process as a pixel classification which is fully based on color. The segmentator operates directly upon the image pixels using the classification properties of the LVQ networks. The algorithm is effectively applied to the segmentation of sampled images showing its capacity to satisfactorily segment color despite remarkable illumination differences, considering indoor and outdoor scenes. The results demonstrate the operation of the LVQ algorithm which in turn is capable of organizing topologically the input space, accomplishing the segmentation process, despite a small number of neurons.

The presented system has two important features. First, since the LVQ algorithm works directly on the image pixels with no dynamic model or probability distribution, the execution time is faster than other approaches. Second, the algorithm exhibits interesting generalization properties, in particular considering images with changing illumination.

Further increases in the segmentator performance might be reached, if the parameter σ is also adapted using some kind of optimization technique.

References

1. de-Ridder D, Handels H (2002) Image processing with neural networks—a review. Pattern Recognit 35:2279–2301
2. Cheng HD, Jiang XH, Sun Y, Wang JL (2001) Color image segmentation: advances and prospects. Pattern Recognit 34(12):2259–2281
3. Gonzalez RC, Woods RE (2000) Digital image processing. Edgard Blücher, São Paulo
4. Park SH, Yun ID, Lee SU (2001) Color image segmentation based on 3D clustering: morphological approach. Pattern Recognit 31(8):1061–1076
5. Chen TQ, Lu Y (2002) Color image segmentation: an innovative approach. Pattern Recognit 35(2):395–405
6. Nikolaev D, Nikolayev P (2004) Linear color segmentation and its implementation. Comput Vis Image Underst 94(3):115–139
7. Ohlander R, Price K, Reddy DR (1980) Picture segmentation using a recursive region splitting method. Comput Graph Image Process 8:313–333
8. Cheng H, Jiang X, Wang J (2002) Color image segmentation based on histogram thresholding and region merging. Pattern Recognit 35(2):373–393
9. Tremeau A, Borel N (1997) A region growing and merging algorithm to color segmentation. Pattern Recognit 30(7):1191–1203
10. Jepson A, Fleet D, El-Maraghi T (2001) Robust online appearance models for visual tracking. Comput Vis Pattern Recognit 25(10):415–422
11. McKenna S, Raja Y, Gong S (1999) Tracking colour objects using adaptive mixture models. Image Vis Comput 17(1):225–231
12. Raja Y, McKenna S, Gong S (1998) Tracking and segmenting people in varying lighting condition using color. In: International conference on face and gesture recognition, pp 228–233

13. Isard M, MacCormick J (2001) BraMBLe: a bayesian multiple-blob tracker". In: International conference on computer vision, pp 34–41
14. Perez P, Hue C, Vermaak J, Gangnet M (2002) Color-based probabilistic tracking. In: European conference on computer vision, pp 661–675
15. Bradski GR (1998) Computer vision face tracking as a component of a perceptual user interface. In: Workshop on applications of computer vision, pp 214–219
16. Dong G, Xie M (2005) Color clustering and learning for image segmentation based on neural networks. IEEE Trans Neural Netw 16(4):925–936
17. Ong SH, Yeo NC, Lee KH, Venkatesh YV, Cao DM (2002) Segmentation of color images using a two-stage self-organizing network. Image Vis Comput 20:279–289
18. Yeo NC, Lee KH, Venkatesh YV, Ong SH (2005) Colour image segmentation using the self-organizing map and adaptive resonance theory. Image Vis Comput 23:1060–1079
19. Comaniciu D, Meer P (1997) Robust analysis of feature spaces: color image segmentation. In: IEEE conference on computer vision and pattern recognition, pp 750–55
20. Pietikainen M (2008) Accurate color discrimination with classification based on feature distributions. In: International conference on pattern recognition, C, pp 833–838
21. Kohonen T (1997) Self-organizing maps, 2nd edn. Springer, Berlin
22. Kohonen T (1987) Self-organization and associative memory, 2nd edn. Springer, Berlin
23. Jang GJ, Kweon IO (2001) Robust real-time face tracking using adaptive color model. In: International conference on robotics and automation, pp 138–149
24. Nummiaro K, Koller-Meier E, Van-Gool L (2002) A color-based particle filter. In: Proceedings of the 1st international workshop on generative-model-based vision, pp 353–358
25. Chiunhsiun L (2007) Face detection in complicated backgrounds and different illumination conditions by using YCbCr color space and neural network. Pattern Recognit Lett 28:2190–2200
26. Salmond D (1990) Mixture reduction algorithms for target tracking in clutter. SPIE Signal Data Process Small Targets 1305:434–445

Chapter 5
Global Optimization Using Opposition-Based Electromagnetism-Like Algorithm

Electromagnetism-like optimization (EMO) is a global optimization algorithm which allows to solve complex multimodal optimization problems. EMO employs search agents that emulate a population of charged particles which interact to each other according to electromagnetism's laws (attraction and repulsion). Traditionally, EMO requires a large number of generations in its local search procedure. If this local search process is eliminated, it is severely damaged the overall convergence, exploration, population diversity and accuracy. This chapter presents an enhanced EMO algorithm called OBEMO, which uses the opposition-based learning (OBL) approach to accelerate the global convergence speed. OBL is a machine intelligence strategy which considers the current candidate solution and its opposite value at the same time, achieving a faster exploration of the search space. The presented OBEMO method significantly reduces the required computational effort without causing any detriment to the search capabilities of the original EMO algorithm. Experiments showed that OBEMO obtains promising performance for most of the discussed test problems.

5.1 Introduction

Global optimization (GO) [1, 2] has issued applications for many areas of science [3], engineering [4], economics [5, 6] and others whose definition requires mathematical modelling [7, 8]. In general, GO aims to find the global optimum for an objective function which has been defined over a given search space. The difficulties associated with the use of mathematical methods over GO problems have contributed to the development of alternative solutions. Linear programming and dynamic programming techniques, for example, often have failed in solving (or reaching local optimum at) NP-hard problems which feature a large number of variables and non-linear objective functions. In order to overcome such problems,

© Springer International Publishing AG 2017 77
M.-A. Díaz-Cortés et al., *Engineering Applications of Soft Computing*,
Intelligent Systems Reference Library 129, DOI 10.1007/978-3-319-57813-2_5

researchers have proposed metaheuristic-based algorithms for searching near-optimum solutions.

Metaheuristic algorithms are stochastic search methods that mimic the metaphor of biological or physical phenomena. The core of such methods lies on the analysis of collective behaviour of relatively simple agents working on decentralized systems. Such systems typically gather an agent's population that can communicate to each other while sharing a common environment. Despite a non-centralized control algorithm regulates the agent behaviour, the agent can solve complex tasks by analyzing a given global model and harvesting cooperation to other elements. Therefore, a novel global behaviour evolves from interaction among agents as it can be seen on typical examples that include ant colonies, animal herding, bird flocking, fish schooling, honey bees, bacteria, charged particles and many more. Some other metaheuristic optimization algorithms have been recently proposed to solve optimization problems, such as genetic algorithms (GA) [9], particle swarm optimization (PSO) [10], ant colony optimization (ACO) [11], differential evolution (DE) [12], Artificial Immune Systems (AIS) [13] and artificial bee colony [14] and gravitational search algorithm (GSA) [15].

Electromagnetism-like algorithm (EMO) is a relatively new population-based meta-heuristic algorithm which was firstly introduced by Birbil and Fang [16] to solve continuous optimization models using bounded variables. The algorithm imitates the attraction–repulsion mechanism between charged particles in an electromagnetic field. Each particle represents a solution and carries a certain amount of charge which is proportional to the solution quality (objective function). In turn, solutions are defined by position vectors which give real positions for particles within a multi-dimensional space. Moreover, objective function values of particles are calculated considering such position vectors. Each particle exerts repulsion or attraction forces over other population members; the resultant force acting over a particle is used to update its position. Clearly, the idea behind the EMO methodology is to move particles towards the optimum solution by exerting attraction or repulsion forces. Unlike other traditional meta-heuristics techniques such as GA, DE, ABC and AIS, whose population members exchange materials or information between each other, the EMO methodology assumes that each particle is influenced by all other particles in the population, mimicking other heuristics methods such as PSO and ACO. Although the EMO algorithm shares some characteristics with PSO and ACO, recent works have exhibited its better accuracy regarding optimal parameters [17–20], yet showing convergence [21]. EMO has been successfully applied to solve different sorts of engineering problems such as flow-shop scheduling [22], communications [23], vehicle routing [24], array pattern optimization in circuits [25], neural network training [26] control systems [27] and image processing [28].

EMO algorithm employs four main phases: initialization, local search, calculation and movement. The local search procedure is a stochastic search in several directions over all coordinates of each particle. EMO's main drawback is its computational complexity resulting from the large number of iterations which are commonly required during the searching process. The issue becomes worst as the dimension of the optimization problem increases. Several approaches, which simplify the local search, have been proposed in the literature to reduce EMO's computational effort. In [29] where Guan et al. proposed a discrete encoding for the particle set in order to reduce search directions at each dimension. In [30, 31], authors include a new local search method which is based on a fixed search pattern and a shrinking strategy that aims to reduce the population size as the iterative process progresses. Additionally, in [17], a modified local search phase that employs the gradient descent method is adopted to enhance its computational complexity. Although all these approaches have improved the computational time which is required by the original EMO algorithm, recent works [27, 32] have demonstrated that reducing or simplifying EMO's local search processes also affects other important properties, such as convergence, exploration, population diversity and accuracy.

On the other hand, the opposition-based learning (OBL), that has been initially proposed in [33], is a machine intelligence strategy which considers the current estimate and its correspondent opposite value (i.e., guess and opposite guess) at the same time to achieve a fast approximation for a current candidate solution. It has been mathematically proved [34–36] that an opposite candidate solution holds a higher probability for approaching the global optimum solution than a given random candidate, yet quicker. Recently, the concept of opposition has been used to accelerate metaheuristic-based algorithms such as GA [37], DE [38], PSO [39] and GSA [40].

In this chapter, an opposition-based EMO called OBEMO is presented, it was constructed by combining the opposition-based strategy and the standard EMO technique. The enhanced algorithm allows a significant reduction on the computational effort which required by the local search procedure yet avoiding any detriment to the good search capabilities and convergence speed of the original EMO algorithm. The presented algorithm has been experimentally tested by means of a comprehensive set of complex benchmark functions. Comparisons to the original EMO and others state-of-the-art EMO-based algorithms [7] demonstrate that the OBEMO technique is faster for all test functions, yet delivering a higher accuracy. Conclusions on the conducted experiments are supported by statistical validation that properly supports the results.

The rest of the chapter is organized as follows: Sect. 5.2 introduces the standard EMO algorithm. Section 5.3 gives a simple description of OBL and Sect. 5.4 explains the implementation of the proposed OBEMO algorithm. Section 5.5 presents a comparative study among OBEMO and other EMO variants over several benchmark problems. Finally, some conclusions are drawn in Sect. 5.6.

5.2 Electromagnetism: Like Optimization Algorithm (EMO)

EMO algorithm is a simple and direct search algorithm which has been inspired by the electro-magnetism phenomenon. It is based on a given population and the optimization of global multi-modal functions. In comparison to GA, it does not use crossover or mutation operators to explore feasible regions; instead it does implement a collective attraction–repulsion mechanism yielding a reduced computational cost with respect to memory allocation and execution time. Moreover, no gradient information is required as it employs a decimal system which clearly contrasts to GA. Few particles are required to reach converge as has been already demonstrated in [11].

EMO algorithm can effectively solve a special class of optimization problems with bounded variables in the form of:

$$\min f(x) \atop x \in [l, u] \tag{5.1}$$

where $[l, u] = \{x \in \Re^n | l_d \leq x_d \leq u_d, d = 1, 2 \ldots n\}$ and n being the dimension of the variable x, $[l, u] \subset \Re^n$, a nonempty subset and a real-value function $f : [l, u] \rightarrow \Re$. Hence, the following problem features are known:

- n: dimensional size of the problem.
- u_d: the highest bound of the kth dimension.
- l_d: the lowest bound of the kth dimension.
- $f(x)$: the function to be minimized.

EMO algorithm has four phases [6]: initialization, local search, computation of the total force vector and movement. A deeper discussion about each stage follows.

Initialization, a number of m particles is gathered as their highest (u) and lowest limit (l) are identified.

Local search, gathers local information for a given point $\mathbf{g}^p$, where $p \in (1, \ldots, m)$.

Calculation of the total force vector, charges and forces are calculated for every particle.

Movement, each particle is displaced accordingly, matching the corresponding force vector.

5.2.1 Initialization

First, the population of m solutions is randomly produced at an initial state. Each n-dimensional solution is regarded as a charged particle holding a uniform distribution between the highest (u) and the lowest (l) limits. The optimum particle (solution) is thus defined by the objective function to be optimized. The procedure ends when all the m samples are evaluated, choosing the sample (particle) that has gathered the best function value.

5.2.2 Local Search

The local search procedure is used to gather local information within the neighbourhood of a candidate solution. It allows obtaining a better exploration and population diversity for the algorithm.

Considering a pre-fixed number of iterations known as *ITER* and a feasible neighbourhood search δ, the procedure iterates as follows: point $\mathbf{g}^p$ is assigned to a temporary point $\mathbf{t}$ to store the initial information. Next, for a given coordinate d, a random number is selected (λ_1) and combined with δ as a step length, which in turn, moves the point $\mathbf{t}$ along the direction d, with a randomly determined sign (λ_2). If point $\mathbf{t}$ observes a better performance over the iteration number *ITER*, point $\mathbf{g}^p$ is replaced by $\mathbf{t}$ and the neighbourhood search for point $\mathbf{g}^p$ finishes, otherwise $\mathbf{g}^p$ is held. The pseudo-code is listed in Fig. 5.1.

In general, the local search for all particles can also reduce the risk of falling into a local solution but is time consuming. Nevertheless, recent works [17, 32] have shown that eliminating, reducing or simplifying the local search process affects significantly the convergence, exploration, population diversity and accuracy of the EMO algorithm.

1:	*Counter* $\leftarrow 1$	12:	$\mathbf{t}_d \leftarrow \mathbf{t}_d - \lambda_2 (Length)$
2:	*Length* $\leftarrow \delta (\max\{u_d - l_d\})$	13:	**end if**
3:	**for** $p = 1$ **to** m **do**	14:	**if** $f(\mathbf{t}) < f(\mathbf{g}^p)$ **then**
4:	**for** $d = 1$ **to** n **do**	15:	$\mathbf{g}^p \leftarrow \mathbf{t}$
5:	$\lambda_1 \leftarrow U(0,1)$	16:	counter ← ITER − 1
6:	**while** *Counter* < *ITER* **do**	17:	**end if**
7:	$\mathbf{t} \leftarrow \mathbf{g}^p$	18:	*Counter* $\leftarrow$ *Counter* $+1$
8:	$\lambda_2 \leftarrow U(0,1)$	19:	**end while**
9:	**if** $\lambda_1 > 0.5$ **then**	20:	**end for**
10:	$\mathbf{t}_d \leftarrow \mathbf{t}_d + \lambda_2 (Length)$	21:	**end for**
11:	**Else**	22:	$\mathbf{g}^{best} \leftarrow \mathrm{argmin}\{f(\mathbf{g}^p), \forall p\}$

Fig. 5.1 Pseudo-code list for the local search algorithm

5.2.3   Total Force Vector Computation

The total force vector computation is based on the *superposition principle* (Fig. 5.2) from the electro-magnetism theory which states: "the force exerted on a point via other points is inversely proportional to the distance between the points and directly proportional to the product of their charges" [41]. The particle moves following the resultant Coulomb's force which has been produced among particles as a charge-like value. In the EMO implementation, the charge for each particle is determined by its fitness value as follows:

$$q^p = \exp\left(-n\frac{f(\mathbf{g}^p) - f(\mathbf{g}^{best})}{\sum\limits_{h=1}^{m}(f(\mathbf{g}^h) - f(\mathbf{g}^{best}))}\right), \forall p \tag{5.2}$$

where n denotes the dimension of $\mathbf{g}^p$ and m represents the population size. A higher dimensional problem usually requires a larger population. In Eq. (5.2), the particle showing the best fitness function value $\mathbf{g}^{best}$ is called the "*best particle*", getting the highest charge and attracting other particles holding high fitness values. The repulsion effect is applied to all other particles exhibiting lower fitness values. Both effects, attraction-repulsion are applied depending on the actual proximity between a given particle and the best-graded element.

The overall resultant force between all particles determines the actual effect of the optimization process. The final force vector for each particle is evaluated under the Coulomb's law and the superposition principle as follows:

$$\mathbf{F}^p = \sum\limits_{h\neq p}^{m} \begin{cases} (\mathbf{g}^h - \mathbf{g}^p)\frac{q^p q^h}{\|\mathbf{g}^h - \mathbf{g}^p\|^2} & if \quad f(\mathbf{g}^h) < f(\mathbf{g}^p) \\ (\mathbf{g}^p - \mathbf{g}^h)\frac{q^p q^h}{\|\mathbf{g}^h - \mathbf{g}^p\|^2} & if \quad f(\mathbf{g}^h) \geq f(\mathbf{g}^p) \end{cases}, \forall p \tag{5.3}$$

where $f(\mathbf{g}^h) < f(\mathbf{g}^p)$ represents the attraction effect and $f(\mathbf{g}^h) \geq f(\mathbf{g}^p)$ represents the repulsion force (see Fig. 5.3). The resultant force of each particle is proportional to the product between charges and is inversely proportional to the distance between

Fig. 5.2 The superposition principle

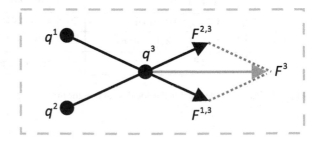

Fig. 5.3 Coulomb law: α represents the distance between charged particles, q^1, q^2 are the charges, and F is the exerted force as has been generated by the charge interaction

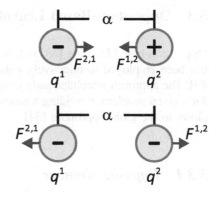

particles. In order to keep feasibility, the vector in expression (5.3) should be normalized as follows:

$$\hat{\mathbf{F}}^p = \frac{\mathbf{F}^p}{\|\mathbf{F}^p\|}, \ \forall p. \tag{5.4}$$

5.2.4 Movement

The change of the d-coordinate for each particle p is computed with respect to the resultant force as follows:

$$g_d^p = \begin{cases} g_d^p + \lambda \cdot \hat{F}_d^p \cdot (u_d - g_d^p) & \text{if} \quad \hat{F}_d^p > 0 \\ g_d^p + \lambda \cdot \hat{F}_d^p \cdot (g_d^p - l_d) & \text{if} \quad \hat{F}_d^p \le 0 \end{cases}, \ \forall p \neq best, \forall d \tag{5.5}$$

In Eq. (5.5), λ is a random step length that is uniformly distributed between zero and one. u_d and l_d represent the upper and lower boundary for the d-coordinate, respectively. $\hat{F}_d^p$ represents the d element of $\hat{\mathbf{F}}^p$. If the resultant force is positive, then the particle moves towards the highest boundary by a random step length. Otherwise it moves toward the lowest boundary. The best particle does not move at all, because it holds the absolute attraction, pulling or repelling all others in the population.

The process is halted when a maximum iteration number is reached or when the value $f(\mathbf{g}^{best})$ is near to zero or to the required optimal value.

5.3 Opposition-Based Learning (OBL)

Opposition-based learning [33] is a new concept in computational intelligence that has been employed to effectively enhance several soft computing algorithms [42, 43]. The approach simultaneously evaluates a solution x and its opposite solution $\bar{x}$ for a given problem, providing a renewed chance to find a candidate solution lying closer to the global optimum [34].

5.3.1 Opposite Number

Let $x \in [l, u]$ be a real number, where l and u are the lowest and highest bound respectively. The opposite of x is defined by:

$$\bar{x} = u + l - x \tag{5.6}$$

5.3.2 Opposite Point

Similarly, the opposite number definition is generalized to higher dimensions as follows: Let $\mathbf{x} = (x_1, x_2, \ldots, x_n)$ be a point within a n-dimensional space, where $x_1, x_2, \ldots, x_n \in R$ and $x_i \in [l_i, u_i]$, $i \in 1, 2, \ldots, n$. The opposite point $\bar{\mathbf{x}} = (\bar{x}_1, \bar{x}_2, \ldots, \bar{x}_n)$ is defined by:

$$\bar{x}_i = u_i + l_i - x_i \tag{5.7}$$

5.3.3 Opposite-Based Optimization

Metaheuristic methods start by considering some initial solutions (initial population) and trying to improve them toward some optimal solution(s). The process of searching ends when some predefined criteria are satisfied. In the absence of a priori information about the solution, random guesses are usually considered. The computation time, among others algorithm characteristics, is related to the distance of these initial guesses taken from the optimal solution. The chance of starting with a closer (fitter) solution can be enhanced by simultaneously checking the opposite solution. By doing so, the fitter one (guess or opposite guess) can be chosen as an initial solution following the fact that, according to probability theory, 50% of the time a guess is further from the solution than its opposite guess [35]. Therefore, starting with the closer of the two guesses (as judged by their fitness values) has the potential to accelerate convergence. The same approach can be applied not only to initial solutions but also to each solution in the current population.

By applying the definition of an opposite point, the opposition-based optimization can be defined as follows: Let $\mathbf{x}$ be a point in a n-dimensional space (i.e. a candidate solution). Assume $f(\mathbf{x})$ is a fitness function which evaluates the quality of such candidate solution. According to the definition of opposite point, $\bar{\mathbf{x}}$ is the opposite of $\mathbf{x}$. If $f(\bar{\mathbf{x}})$ is better than $f(\mathbf{x})$, then $\mathbf{x}$ is updated with $\bar{\mathbf{x}}$, otherwise current point $\mathbf{x}$ is kept. Hence, the best point ($\mathbf{x}$ or $\bar{\mathbf{x}}$) is modified using known operators from the population-based algorithm.

Figure 5.4 shows the opposition-based optimization procedure. In the example, Fig. 5.4a, b represent the function to be optimized and its corresponding contour plot, respectively. By applying the OBL principles to the current population P (see Fig. 5.4b), the three particles $\mathbf{x}_1$, $\mathbf{x}_2$ and $\mathbf{x}_3$ produce a new population OP, gathering particles $\bar{\mathbf{x}}_1$, $\bar{\mathbf{x}}_2$ and $\bar{\mathbf{x}}_3$. The three fittest particles from P and OP are selected as the new population P'. It can be seen from Fig. 5.4b that $\mathbf{x}_1$, $\bar{\mathbf{x}}_2$ and $\bar{\mathbf{x}}_3$ are three new members in P'. In this case, the transformation conducted on $\mathbf{x}_1$ did not provide a best chance of finding a candidate solution closer to the global optimum. Considering the OBL selection mechanism, $\bar{\mathbf{x}}_1$ is eliminated from the next generation.

5.4 Opposition-Based Electromagnetism-Like Optimization Algorithm

Similarly to all metaheuristic-based optimization algorithms, two steps are fundamental for the EMO algorithm: the population initialization and the production of new generations by evolutionary operators. In the approach, the OBL scheme is incorporated to enhance both steps. However, the original EMO is considered as the main algorithm while the opposition procedures are embedded into EMO aiming to accelerate its convergence speed. Figure 5.5 shows a data flow comparison between the EMO and the OBEMO algorithm. The novel extended opposition procedures are explained in the following subsections.

5.4.1 Opposition-Based Population Initialization

In population-based meta-heuristic techniques, the random number generation is the common choice to create an initial population in absence of a priori knowledge. Therefore, as mentioned in Sect. 5.3, it is possible to obtain fitter starting candidate solutions by utilizing OBL despite no a priori knowledge about the solution(s) is available. The following steps explain the overall procedure.

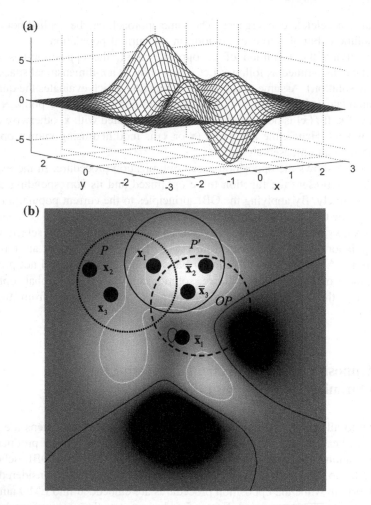

Fig. 5.4 The opposition-based optimization procedure: **a** function to be optimized and **b** its *contour plot*. The current population P includes particles $\mathbf{x}_1$, $\mathbf{x}_2$ and $\mathbf{x}_3$. The corresponding opposite population OP is represented by $\mathbf{x}_1$, $\bar{\mathbf{x}}_2$ and $\bar{\mathbf{x}}_3$. The final population P' is obtained by the OBL selection mechanism yielding particles $\mathbf{x}_1$, $\bar{\mathbf{x}}_2$ and $\bar{\mathbf{x}}_3$

(1) Initialize the population $\mathbf{X}$ with N_p representing the number of particles.
(2) Calculate the opposite population by

$$\bar{x}_i^j = u_i + l_i - x_i^j$$
$$i = 1, 2, \ldots, n; j = 1, 2, \ldots, N_P$$

$$(5.8)$$

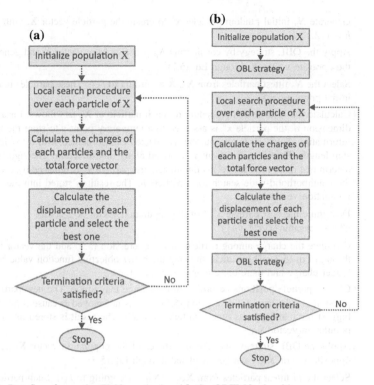

Fig. 5.5 Dataflow for: **a** the EMO method and **b** the OBEMO algorithm

where x_i^j and $\bar{x}_i^j$ denote the ith parameter of the jth particle of the population and its corresponding opposite particle.

(3) Select the N_P fittest elements from $\{\mathbf{X} \cup \bar{\mathbf{X}}\}$ as initial population.

5.4.2   Opposition-Based Production for New Generation

Starting from the current population, the OBL strategy can be used again to produce new populations. In this procedure, the opposite population is calculated and the fittest individuals are selected from the union of the current population and the opposite population. The following steps summarize the OBEMO implementation as follows:

Step 1	Generate N_p initial random particles $\mathbf{x}^h$ to create the particle vector $\mathbf{X}$, with $h \in 1, 2, \ldots N_p$
Step 2	Apply the OBL strategy by considering N_p particles from vector $\mathbf{X}$ and generating the opposite vector $\overline{\mathbf{X}}$ through Eq. (5.7)
Step 3	Select the N_p fittest particles from $\mathbf{X} \cup \overline{\mathbf{X}}$ according to $f(\cdot)$. These particles build the initial population $\mathbf{X}_0$
Step 4	Calculate the local search procedure for each particle of $\mathbf{X}_0$ as follows: for a given dimension d, the particle $\mathbf{x}^h$ is assigned to a temporary point y to store the initial information. Next, a random number is selected and combined with δ to yield the step length. Therefore, the point y is moved along that direction. The sign is determined randomly. If $f(\mathbf{x}^h)$ is minimized, the particle $\mathbf{x}^h$ is replaced by y, ending the neighborhood-wide search for a particle h. The result is stored into the population vector $\mathbf{X}_{Local}$
Step 5	Determine the best particle $\mathbf{x}^{best}$ of the population vector $\mathbf{X}_{Local}$(with $\mathbf{x}^{best} \leftarrow \arg\min\{f(\mathbf{x}^h), \forall h\}$)
Step 6	Calculate the charge among particles using expression (5.2) and the vector force through Eq. (5.3). The particle showing the better objective function value holds a bigger charge and therefore a bigger attraction force
Step 7	Change particle positions according to their force magnitude. The new particle's position is calculated by expression (5.5). $\mathbf{x}^{best}$ is not moved because it has the biggest force and attracts others particles to itself. The result is stored into the population vector $\mathbf{X}_{Mov}$
Step 8	Apply the OBL strategy over the m particles of the population vector $\mathbf{X}_{Mov}$, the opposite vector $\overline{\mathbf{X}}_{Mov}$ can be calculated through Eq. (5.7)
Step 9	Select the m fittest particles from $\mathbf{X}_{Mov} \cup \overline{\mathbf{X}}_{Mov}$ according to $f(\cdot)$. Such particles represent the population $\mathbf{X}_0$
Step 10	Increase the *Iteration* index. If *iteration = MAXITER or the* value of $f(X)$ is smaller than the pre-defined threshold value, then the algorithm is stopped and the flow jumps to step 11. Otherwise, it jumps to step 4
Step 11	The best particle $\mathbf{x}^{best}$ is selected from the last iteration as it is considered as the solution

5.5 Experimental Results

In order to test the algorithm's performance, the proposed OBEMO is compared to the standard EMO and others state-of-the-art EMO-based algorithms. In this section, the experimental results are discussed in the following subsections:

(5.5.1). Test problems
(5.5.2). Parameter settings for the involved EMO algorithms
(5.5.3). Results and discussions

5.5.1 Test Problems

A comprehensive set of benchmark problems, that includes 14 different global optimization tests, has been chosen for the experimental study. According to their use in the performance analysis, the functions are divided in two different sets: original test functions $(f_1 - f_9)$ and multidimensional functions $(f_{10} - f_{14})$. Every function at this paper is considered as a minimization problem itself.

The original test functions, which are shown in Table 5.1, agree to the set of numerical benchmark functions presented by the original EMO paper at [16]. Considering that such function set is also employed by a vast majority of EMO-based new approaches, its use in our experimental study facilitates its comparison to similar works. More details can be found in [44].

The major challenge of an EMO-based approach is to avoid the computational complexity that arises from the large number of iterations which are required during the local search process. Since the computational complexity depends on the dimension of the optimization problem, one set of multidimensional functions (see Table 5.2) is used in order to assess the convergence and accuracy for each algorithm. Multidimensional functions include a set of five different functions whose dimension has been fixed to 30.

5.5.2 Parameter Settings for the Involved EMO Algorithms

The experimental set aims to compare four EMO-based algorithms including the proposed OBEMO. All algorithms face 14 benchmark problems. The algorithms are listed below:

- Standard EMO algorithm [16];
- Hybridizing EMO with descent search (HEMO) [17];
- EMO with fixed search pattern (FEMO) [30];
- The proposed approach OBEMO.

For the original EMO algorithm described in [16] and the proposed OBEMO, the parameter set is configured considering: $\delta = 0.001$ and $LISTER = 4$. For the HEMO, the following experimental parameters are considered:$LsIt_{\max} = 10$, $\varepsilon_r = 0.001$ and $\gamma = 0.00001$. Such values can be assumed as the best configuration set according to [17]. Diverging from the standard EMO and the OBEMO algorithm, the HEMO method reduces the local search phase by only processing the best found particle $\mathbf{x}^{best}$. The parameter set for the FEMO approach is defined by considering the following values: $N_{fe}^{\max} = 100$, $N_{ls}^{\max} = 10$, $\delta = 0.001$, $\delta^{\min} = 1 \times 10^{-8}$ and $\varepsilon_\delta = 0.1$. All aforementioned EMO-based algorithms use the same population size of $m = 50$.

Table 5.1 Optimization test functions corresponding to the original test set ($f_1 - f_9$)

Function	Search domain	Global minima
Branin $f_1(x_1,x_2) = (x_2 - \frac{5}{4\pi^2}x_1^2 + \frac{5}{\pi}x_1 - 6)^2 + 10(1 - \frac{1}{8\pi})\cos x_1 + 10$	$-5 \leq x_1 \leq 100 \leq x_2 \leq 15$	0.397887
Camel $f_2(x_1,x_2) = -\dfrac{-x_1^2 + 4.5x_2^2 + 2}{e^{2x_2^2}}$	$-2 \leq x_1, x_2 \leq 2$	−1.031
Goldenstain-price $f_3(x_1,x_2) = 1 + (x_1 + x_2 + 1)^2 \times (19 - 14x_1 + 13x_1^2 - 14x_2 + 6x_1x_2 + 3x_2^2)$ $\times (30 + 2x_1 - 3x_2)^2 \times (18 - 32x_1 + 12x_1^2 - 48x_2 - 36x_1x_2 + 27x_2^2)$	$-2 \leq x_1, x_2 \leq 2$	3.0
Hartmann (3-dimensional) $f_4(\mathbf{x}) = -\sum\limits_{i=1}^{4} \alpha_i \exp\left[-\sum\limits_{j=1}^{3} A_{ij}(x_j - P_{ij})^2\right]$ $\alpha = [1, 1.2, 3, 3.2]$, $\mathbf{A} = \begin{bmatrix} 3.0 & 10 & 30 \\ 0.1 & 10 & 35 \\ 3.0 & 10 & 35 \\ 3.0 & 10 & 35 \end{bmatrix}$, $\mathbf{P} = 10^{-4}\begin{bmatrix} 6890 & 1170 & 2673 \\ 4699 & 4387 & 7470 \\ 1091 & 8732 & 5547 \\ 381 & 5743 & 8828 \end{bmatrix}$	$0 \leq x_i \leq 1$ $i = 1, 2, 3$	−3.8627
Hartmann (6-dimensional) $f_5(\mathbf{x}) = -\sum\limits_{i=1}^{4} \alpha_i \exp\left[-\sum\limits_{j=1}^{6} B_{ij}(x_j - Q_{ij})^2\right]$ $\alpha = [1, 1.2, 3, 3.2]$, $\mathbf{B} = \begin{bmatrix} 10 & 3 & 17 & 3.05 & 1.7 & 8 \\ 0.05 & 10 & 17 & 0.1 & 8 & 14 \\ 3 & 3.5 & 1.7 & 10 & 17 & 8 \\ 17 & 8 & 0.05 & 10 & 0.1 & 14 \end{bmatrix}$, $\mathbf{Q} = 10^{-4}\begin{bmatrix} 1312 & 1696 & 5569 & 124 & 8283 & 5886 \\ 2329 & 4135 & 8307 & 3736 & 1004 & 9991 \\ 2348 & 1451 & 3522 & 2883 & 3047 & 6650 \\ 4047 & 8828 & 8732 & 5743 & 1091 & 381 \end{bmatrix}$	$0 \leq x_i \leq 1 i = 1, 2, 3, \ldots, 6$	−3.8623

(continued)

Table 5.1 (continued)

Function	Search domain	Global minima
Shekel S_m (4-dimensional) $$S_m(\mathbf{x}) = -\sum_{j=1}^{m}\left[\sum_{i=1}^{4}(x_i - C_{ij})^2 + \beta_j\right]^{-1}$$ $\boldsymbol{\beta} = [1,2,2,4,4,6,3,7,5,5]^T$, $$\mathbf{C} = \begin{bmatrix} 4.0 & 1.0 & 8.0 & 6.0 & 3.0 & 2.0 & 5.0 & 8.0 & 6.0 & 7.0 \\ 4.0 & 1.0 & 8.0 & 6.0 & 7.0 & 9.0 & 5.0 & 1.0 & 2.0 & 3.6 \\ 4.0 & 1.0 & 8.0 & 6.0 & 3.0 & 2.0 & 3.0 & 8.0 & 6.0 & 7.0 \\ 4.0 & 1.0 & 8.0 & 6.0 & 7.0 & 9.0 & 3.0 & 1.0 & 2.0 & 3.6 \end{bmatrix}$$	$0 \le x_i \le 1$ $i = 1,2,3,4$	
$f_6(\mathbf{x}) = S_5(\mathbf{x})$		-10.1532
$f_7(\mathbf{x}) = S_7(\mathbf{x})$		-10.4029
$f_8(\mathbf{x}) = S_{10}(\mathbf{x})$		-10.5364
Shubert $f_9(x_1,x_2) = \left(\sum_{i=1}^{5} i\cos((i+1)x_1 + i)\right)\left(\sum_{i=1}^{5} i\cos((i+1)x_2 + i)\right)$	$-10 \le x_1, x_2 \le 10$	-186.73

Table 5.2 Multidimensional test function set ($f_{10} - f_{14}$)

Function	Search domain	Global minima
$f_{10}(\mathbf{x}) = \sum_{i=1}^n \left[x_i^2 - 10 \cos(2\pi x_i) + 10 \right]$	$[-5.12, 5.12]^{30}$	0
$f_{11}(\mathbf{x}) = -20 \exp\left(-0.2\sqrt{\frac{1}{n}\sum_{i=1}^n x_i^2}\right) - \exp\left(\frac{1}{n}\sum_{i=1}^n \cos(2\pi x_i)\right) + 20$	$[-32, 32]^{30}$	0
$f_{12}(\mathbf{x}) = \frac{1}{4000}\sum_{i=1}^n x_i^2 - \prod_{i=1}^n \cos\left(\frac{x_i}{\sqrt{i}}\right) + 1$	$[-600, 600]^{30}$	0
$f_{13}(\mathbf{x}) = \frac{\pi}{n}\left\{ 10 \sin(\pi y_1) + \sum_{i=1}^{n-1} (y_i - 1)^2 \left[1 + 10 \sin^2(\pi y_{i+1}) \right] + (y_n - 1)^2 \right\}$ $+ \sum_{i=1}^n u(x_i, 10, 100, 4)$ $y_i = 1 + \frac{x_i + 1}{4} \quad u(x_i, a, k, m) = \begin{cases} k(x_i - a)^m & x_i > a \\ 0 & -a < x_i < a \\ k(-x_i - a)^m & x_i < -a \end{cases}$	$[-50, 50]^{30}$	0
$f_{14}(\mathbf{x}) = \sin^2(3\pi x_1) + \sum_{i=1}^n (x_i - 1)^2 \left[1 + \sin^2(3\pi x_i + 1) \right]$ $+ (x_n - 1)^2 \left[1 + \sin^2(2\pi x_n) \right] + \sum_{i=1}^n u(x_i, 5, 100, 4)$	$[-50, 50]^{30}$	0

5.5.3 Results

Original test functions set

On this test set, the performance of the OBEMO algorithm is compared to standard EMO, HEMO and FEMO, considering the original test functions set. Such functions, presented in Table 5.1, hold different dimensions and one known global minimum. The performance is analysed by considering 35 different executions for each algorithm. The case of no significant changes in the solution being registered (i.e. smaller than 10^{-4}) is considered as stopping criterion.

The results, shown by Table 5.3, are evaluated assuming the averaged best value $f(x)$ and the average number of executed iterations (*MAXITER*). Figure 5.6 shows the optimization process for the function f_3 and f_6. Such function values correspond to the best case for each approach that is obtained after 35 executions.

In order to statistically analyse the results in Table 5.3, a non-parametric significance proof known as the Wilcoxon's rank test [45–47] has been conducted. Such proof allows assessing result differences among two related methods. The analysis is performed considering a 5% significance level over the "averaged best value of $f(x)$" and the "averaged number of executed iterations of *MAXITER*" data. Tables 5.4 and 5.5 reports the p-values produced by Wilcoxon's test for the pair-wise comparison of the "averaged best value" and the "averaged number of executed iterations" respectively, considering three groups. Such groups are formed by OBEMO versus EMO, OBEMO versus HEMO and OBEMO versus FEMO. As a null hypothesis, it is assumed that there is no difference between the values of the two algorithms. The alternative hypothesis considers an actual difference between values from both approaches. The results obtained by the Wilcoxon test indicate that data cannot be assumed as occurring by coincidence (i.e. due to the normal noise contained in the process).

Table 5.4 considers the Wilcoxon analysis with respect to the "averaged best value" of $f(x)$. The p-values for the case of OBEMO versus EMO are larger than 0.05 (5% significance level) which is a strong evidence supporting the null hypothesis which indicates that there is no significant difference between both methods. On the other hand, in cases for the p-values corresponding to the OBEMO versus HEMO and OBEMO versus FEMO, they are less than 0.05 (5% significance level), which accounts for a significant difference between the "averaged best value" data among methods. Table 5.5 considers the Wilcoxon analysis with respect to the "averaged number of executed iterations" values. Applying the same criteria, it is evident that there is a significant difference between the OBEMO versus EMO case, despite the OBEMO versus HEMO and OBEMO versus FEMO cases offering similar results.

Multidimensional functions

In contrast to the original functions, Multidimensional functions exhibit many local minima/maxima which are, in general, more difficult to optimize. In this section the performance of the OBEMO algorithm is compared to the EMO, the HEMO and the

Table 5.3 Comparative results for the EMO, the OBEMO, the HEMO and the FEMO algorithms considering the original test functions set (Table 5.1)

Function	f_1	f_2	f_3	f_4	f_5	f_6	f_7	f_8	f_9
Dimension	2	2	2	3	6	4	4	4	2
EMO									
Averaged best values $f(x)$	0.3980	−1.015	3.0123	−3.7156	−3.6322	−10.07	−10.23	−10.47	−186.71
Averaged *MAXITER*	103	128	197	1.59E + 03	1.08E + 03	30	31	29	44
OBEMO									
Averaged best values $f(x)$	0.3980	−1.027	3.0130	−3.7821	−3.8121	−10.11	−10.22	−10.50	−186.65
Averaged *MAXITER*	61	83	101	1.12E + 03	826	18	19	17	21
HEMO									
Averaged best values $f(x)$	0.5151	−0.872	3.413	−3.1187	−3.0632	−9.041	−9.22	−9.1068	−184.31
Averaged *MAXITER*	58	79	105	1.10E + 03	805	17	18	15	22
FEMO									
Averaged best values $f(x)$	0.4189	−0.913	3.337	−3.3995	−3.2276	−9.229	−9.88	−10.18	−183.88
Averaged *MAXITER*	63	88	98	1.11E + 03	841	21	22	19	25

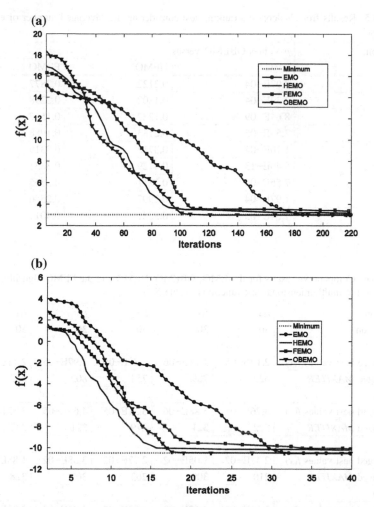

Fig. 5.6 Comparison of the optimization process for two original test functions: **a** f_3 and **b** f_6

Table 5.4 Results from Wilcoxon's ranking test considering the "averaged best value of $f(x)$"

Function	p-Values OBEMO versus		
	EMO	HEMO	FEMO
f_1	0.3521	1.21E−04	1.02E−04
f_2	0.4237	1.05E−04	0.88E−04
f_3	0.2189	4.84E−05	3.12E−05
f_4	0.4321	1.35E−05	1.09E−05
f_5	0.5281	2.73E−04	2.21E−04
f_6	0.4219	1.07E−04	0.77E−04
f_7	0.3281	3.12E−05	2.45E−05
f_8	0.4209	4.01E−05	3.62E−05
f_9	0.2135	1.86E−05	1.29E−05

Table 5.5 Results from Wilcoxon's ranking test considering the "averaged number of executed iterations"

Function	p-Values OBEMO versus		
	EMO	HEMO	FEMO
f_1	2.97E−04	0.2122	0.2877
f_2	3.39E−04	0.1802	0.2298
f_3	8.64E−09	0.1222	0.1567
f_4	7.54E−05	0.2183	0.1988
f_5	1.70E−04	0.3712	0.3319
f_6	5.40E−13	0.4129	0.3831
f_7	7.56E−04	0.3211	0.3565
f_8	1.97E−04	0.2997	0.2586
f_9	1.34E−05	0.3521	0.4011

Table 5.6 Comparative results for the EMO, OBEMO, HEMO and the FEMO algorithms being applied to the multidimensional test functions (Table 5.2)

Function	f_{10}	f_{11}	f_{12}	f_{13}	f_{14}
Dimension	30	30	30	30	30
EMO					
Averaged best values $f(x)$	2.12E−05	1.21E−06	1.87E−05	1.97E−05	2.11E−06
Averaged *MAXITER*	622	789	754	802	833
OBEM					
Averaged best values $f(x)$	3.76E−05	5.88E−06	3.31E−05	4.63E−05	3.331E−06
Averaged *MAXITER*	222	321	279	321	342
HEMO					
Averaged best values $f(x)$	2.47E−02	1.05E−02	2.77E−02	3.08E−02	1.88E−2
Averaged *MAXITER*	210	309	263	307	328
FEMO					
Averaged best values $f(x)$	1.36E−02	2.62E−02	1.93E−02	2.75E−02	2.33E−02
Averaged *MAXITER*	241	361	294	318	353

FEMO algorithms, considering functions in Table 5.2. This comparison reflects the algorithm's ability to escape from poor local optima and to locate a near-global optimum, consuming the least number of iterations. The dimension of such functions is set to 30. The results (Table 5.6) are averaged over 35 runs reporting the "averaged best value" and the "averaged number of executed iterations" as performance indexes.

The Wilcoxon rank test results, presented in Table 5.7, shows that the p-values (regarding to the "averaged best value" values of Table 5.6) for the case of OBEMO versus EMO, indicating that there is no significant difference between both methods. p-values corresponding to the OBEMO versus HEMO and OBEMO versus FEMO show that there is a significant difference between the "averaged

Table 5.7 Results from Wilcoxon's ranking test considering the "best averaged values"

Function	p-Values OBEMO versus		
	EMO	HEMO	FEMO
f_{10}	0.2132	3.21E−05	3.14E−05
f_{11}	0.3161	2.39E−05	2.77E−05
f_{12}	0.4192	5.11E−05	1.23E−05
f_{13}	0.3328	3.33E−05	3.21E−05
f_{14}	0.4210	4.61E−05	1.88E−05

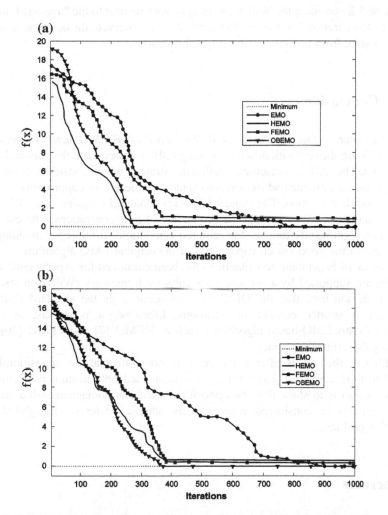

Fig. 5.7 Optimization process comparison for two multidimensional test functions: **a** f_{12} and **b** f_{14}

best" values among the methods. Figure 5.7 shows the optimization process for the function f_{12} and f_{14}. Such function values correspond to the best case, for each approach, obtained after 35 executions.

Table 5.8 Results from Wilcoxon's ranking test considering the "averaged number of executed iterations"

Function	p-Values OBEMO versus		
	EMO	HEMO	FEMO
f_{10}	3.78E−05	0.1322	0.2356
f_{11}	2.55E−05	0.2461	0.1492
f_{12}	6.72E−05	0.3351	0.3147
f_{13}	4.27E−05	0.2792	0.2735
f_{14}	3.45E−05	0.3248	0.3811

Table 5.8 considers the Wilcoxon analysis with respect to the "averaged number of executed iterations" values of Table 5.6. As it is observed, the outcome is similar to the results from last test on the original functions.

5.6   Conclusions

In this chapter, an Opposition-Based EMO, named as OBEMO, has been presented by combining the opposition-based learning (OBL) strategy and the standard EMO technique. The OBL is a machine intelligence strategy which considers, at the same time, a current estimate and its opposite value to achieve a fast approximation for a given candidate solution. The standard EMO is enhanced by using two OBL steps: the population initialization and the production of new generations. The enhanced algorithm significantly reduces the required computational effort yet avoiding any detriment to the good search capabilities of the original EMO algorithm.

A set of 14 benchmark test functions has been employed for experimental study. Results are supported by a statistically significant framework (Wilcoxon test [45–47]) to demonstrate that the OBEMO is as accurate as the standard EMO yet requiring a shorter number of iterations. Likewise, it is as fast as others state-of-the-art EMO-based algorithms such as HEMO [7] and FEMO [30], still keeping the original accuracy.

Although the results offer evidence to demonstrate that the opposition-based EMO method can yield good results on complicated optimization problems, the chapter's aim is to show that the opposition-based electromagnetism-like method can effectively be considered as an attractive alternative for solving global optimization problems.

References

1. Tan S, Cheng X, Hongbo X (2007) An efficient global optimization approach for rough set based dimensionality reduction. Int J Innov Comput Inf Control 3(3):725–736
2. Borji A, Hamidi M (2009) A new approach to global optimization motivated by parliamentary political competitions. Int J Innov Comput Inf Control 5(6):1643–1653

3. Yang C-N, Huang K-S, Yang C-B, Hsu C-Y (2009) Error-tolerant minimum finding with DNA computing. Int J Innov Comput Inf Control 5(10(A)):3045–3057
4. Gao W, Ren H (2011) An optimization model based decision support system for distributed energy systems planning. Int J Innov Comput Inf Control 7(5(B)):2651–2668
5. Chunhui X, Wang J, Shiba N (2007) Multistage portfolio optimization with var as risk measure. Int J Innov Comput Inf Control 3(3):709–724
6. Chang J-F (2009) A performance comparison between genetic algorithms and particle swarm optimization applied in constructing equity portfolios. Int J Innov Comput Inf Control 5(12 (B)):5069–5079
7. Takeuchi Y (2008) Optimization of linear observations for the stationary kalman filter based on a generalized water filling theorem. Int J Innov Comput Inf Control 4(1):211–230
8. Borzabadi AH, Sadjadi ME, Moshiri B (2010) A numerical scheme for approximate optimal control of nonlinear hybrid systems. Int J Innov Comput Inf Control 6(6):2715–2724
9. Holland JH (1975) Adaptation in natural and artificial systems. University of Michigan Press, Ann Arbor
10. Kennedy J, Eberhart R (1995) Particle swarm optimization. In: IEEE international conference on neural networks, Piscataway, NJ, pp 1942–1948
11. Dorigo M, Maniezzo V, Colorni A (1991) Positive feedback as a search strategy, Technical Report 91-016, Politecnico di Milano, Italy
12. Price K, Storn R, Lampinen A (2005) Differential evolution a practical approach to global optimization, springer natural computing series. Springer, Berlin
13. Fyfe C, Jain L (2005) Teams of intelligent agents which learn using artificial immune systems. J Netw Comput Appl 29(2–3):147–159
14. Karaboga D (2005) An idea based on honey bee swarm for numerical optimization, technical report-TR06, Erciyes University, Engineering Faculty, Computer Engineering Department
15. Rashedia E, Nezamabadi-pour H, Saryazdi S (2011) Filter modeling using gravitational search algorithm. Eng Appl Artif Intell 24(1):117–122
16. Birbil SI, Fang S-C (2003) An electromagnetism-like mechanism for global optimization. J Global Optim 25:263–282
17. Rocha A, Fernandes E (2009) Hybridizing the electromagnetism-like algorithm with descent search for solving engineering design problems. Int J Comput Math 86:1932–1946
18. Rocha A, Fernandes E (2009) Modified movement force vector in an electromagnetism-like mechanism for global optimization. Optim Methods Softw 24:253–270
19. Tsou CS, Kao CH (2008) Multi-objective inventory control using electromagnetism-like metaheuristic. Int J Prod Res 46:3859–3874
20. Wu P, Wen-Hung Y, Nai-Chieh W (2004) An electromagnetism algorithm of neural network analysis an application to textile retail operation. J Chin Inst Ind Eng 21:59–67
21. Birbil SI, Fang SC, Sheu RL (2004) On the convergence of a population-based global optimization algorithm. J Global Optim 30(2):301–318
22. Naderi B, Tavakkoli-Moghaddam R, Khalili M (2010) Electromagnetism-like mechanism and simulated annealing algorithms for flowshop scheduling problems minimizing the total weighted tardiness and makespan. Knowl Based Syst 23:77–85
23. Hung H-L, Huang Y-F (2011) Peak to average power ratio reduction of multicarrier transmission systems using electromagnetism-like method. Int J Innov Comput Inf Control 7 (5(A)):2037–2050
24. Yurtkuran A, Emel E (2010) A new hybrid electromagnetism-like algorithm for capacitated vehicle routing problems. Expert Syst Appl 37:3427–3433
25. Jhen-Yan J, Kun-Chou L (2009) Array pattern optimization using electromagnetism-like algorithm. AEU Int J Electron Commun 63:491–496
26. Wu P, Wen-Hung Y, Nai-Chieh W (2004) An electromagnetism algorithm of neural network analysis an application to textile retail operation. J Chin Inst Ind Eng 21:59–67
27. Lee CH, Chang FK (2010) Fractional-order PID controller optimization via improved electromagnetism-like algorithm. Expert Syst Appl 37:8871–8878

28. Cuevas E, Oliva D, Zaldivar D, Pérez-Cisneros M, Sossa H (2012) Circle detection using electro-magnetism optimization. Inf Sci 182(1):40–55
29. Guan X, Dai X, Li J (2011) Revised electromagnetism-like mechanism for flow path design of unidirectional AGV systems. Int J Prod Res 49(2):401–429
30. Rocha AMAC, Fernandes E (2007) Numerical experiments with a population shrinking strategy within a electromagnetism-like algorithm. J Math Comput Simul 1(3):238–243
31. Rocha AMAC, Fernandes E (2011) Numerical study of augmented Lagrangian algorithms for constrained global optimization. Optimization 60(10–11):1359–1378
32. Lee C-H, Chang F-K, Kuo C-T, Chang H-H (2012) A hybrid of electromagnetism-like mechanism and back-propagation algorithms for recurrent neural fuzzy systems design. Int J Syst Sci 43(2):231–247
33. Tizhoosh HR (2005) Opposition-based learning: a new scheme for machine intelligence. In: Proceedings of international conference on computational intelligence for modeling control and automation, pp 695–701
34. Rahnamayn S, Tizhoosh HR, Salama M (2007) A novel population initialization method for accelerating evolutionary algorithms. Comput Math Appl 53(10):1605–1614
35. Rahnamayan S, Tizhoosh HR, Salama MMA (2008) Opposition versus randomness in soft computing techniques. Elsevier J Appl Soft Comput 8:906–918
36. Wang Hui, Zhijian Wu, Rahnamayan Shahryar (2010) Enhanced opposition-based differential evolution for solving high-dimensional continuous optimization problems. Soft Comput. doi:10.1007/s00500-010-0642-7
37. Iqbal MA, Khan NK, Multaba H, Rauf Baig A (2011) A novel function optimization approach using opposition based genetic algorithm with gene excitation. Int J Innov Comput Inf Control 7(7(B)):4263–4276
38. Rahnamayan S, Tizhoosh HR, Salama MMA (2008) Opposition-based differential evolution. IEEE Trans Evol Comput 12(1):64–79
39. Wanga H, Wua Z, Rahnamayan S, Liu Y, Ventresca M (2011) Enhancing particle swarm optimization using generalized opposition-based learning. Inf Sci 181:4699–4714
40. Shaw B, Mukherjee V, Ghoshal SP (2012) A novel opposition-based gravitational search algorithm for combined economic and emission dispatch problems of power systems. Electr Power Energy Syst 35:21–33
41. Cowan EW (1968) Basic electromagnetism. Academic Press, New York
42. Tizhoosh HR (2006) Opposition-based reinforcement learning. J Adv Comput Intell Intell Inf 10(3):78–85
43. Shokri M, Tizhoosh HR, Kamel M (2006) Opposition-based Q(k) algorithm. In: Proceedings of IEEE world congress on computational intelligence, pp 646–53
44. Dixon LCW, Szegö GP (1978) The global optimization problem: an introduction, Towards global optimization 2. North-Holland, Amsterdam, pp 1–15
45. Wilcoxon F (1945) Individual comparisons by ranking methods. Biometrics 1:80–83
46. Garcia S, Molina D, Lozano M, Herrera F (2008) A study on the use of non-parametric tests for analyzing the evolutionary algorithms' behaviour: a case study on the CEC'2005 Special session on real parameter optimization. J Heurist. doi:10.1007/s10732-008-9080-4
47. Santamaría J, Cordón O, Damas S, García-Torres JM, Quirin A (2008) Performance evaluation of memetic approaches in 3D reconstruction of forensic objects. Soft Comput. doi:10.1007/s00500-008-0351-7 (in press)

Chapter 6
Multi-threshold Segmentation Using Learning Automata

Multi-threshold selection for image segmentation is considered as a critical pre-processing step for image analysis, pattern recognition and computer vision. This chapter explores the use of the Learning Automata (LA) algorithm to compute the thresholding points for segmentation proposes. LA is a heuristic method which is able to solve complex optimization problems with interesting results in parameter estimation. Different to other optimization approaches, LA explores in the probability space providing appropriate convergence properties and robustness. In this chapter the segmentation task is considered as an optimization problem and the LA is used to generate the image multi-threshold points. In this approach, one 1-D histogram of a given image is approximated through a Gaussian mixture model whose parameters are calculated using the LA algorithm. Each Gaussian function approximating the histogram represents a pixel class and therefore a thresholding point. Experimental results show fast convergence of the method, avoiding the typical sensitivity to initial conditions.

6.1 Introduction

Several image processing applications aim to detect and mark relevant features which may be later analyzed to perform several high-level tasks. In particular, image segmentation seeks to group pixels within meaningful regions. Commonly, gray levels belonging to the object, are substantially different from those featuring the background. Thresholding is thus a simple but effective tool to isolate objects of interest; its applications include several classics such as document image analysis, whose goal is to extract printed characters [1, 2], logos, graphical content, or musical scores; also it is used for map processing which aims to locate lines, legends, and characters [3]. Moreover, it is employed for scene processing, seeking for object detection, marking [4] and for quality inspection of materials [5, 6].

© Springer International Publishing AG 2017   101
M.-A. Díaz-Cortés et al., *Engineering Applications of Soft Computing*,
Intelligent Systems Reference Library 129, DOI 10.1007/978-3-319-57813-2_6

Thresholding selection techniques can be classified into two categories: bi-level and multi-level. In the former, one limit value is chosen to segment an image into two classes: one representing the object and the other one segmenting the background. When distinct objects are depicted within a given scene, multiple threshold values have to be selected for proper segmentation, which is commonly called multilevel thresholding.

A variety of thresholding approaches have been proposed for image segmentation, including conventional methods [7–10] and intelligent techniques (see for instance [11, 12]). Extending the segmentation algorithms to a multilevel approach may cause some inconveniences: (i) they may have no systematic or analytic solution when the number of classes to be detected increases and (ii) they may also show a slow convergence and/or high computational cost [11].

In this chapter, the segmentation algorithm is based on a parametric model holding a probability density function of gray levels which groups a mixture of several Gaussian density functions (Gaussian mixture). Mixtures represent a flexible method of statistical modelling as they are employed in a wide variety of contexts [13]. Gaussian mixture has received considerable attention in the development of segmentation algorithms despite its performance is influenced by the shape of the image histogram and the accuracy of the estimated model parameters [14]. The associated parameters can be calculated considering an approximated maximum a posterior estimation (MAP) or the maximum likelihood (ML) estimation, considering the Expectation Maximization (EM) algorithm [15, 16] or Gradient-based methods [17].

The EM algorithm provides a simple alternative procedure for computing posterior density or likelihood functions. However, its slow convergence speed has been pointed out as the most serious practical problem [18]. When the EM is used aside Gaussian mixtures, the convergence speed depends on the separation of component populations within such mixture. Therefore, the EM algorithms are very sensitive to the choice of the initial values [19]. Moreover, the EM algorithm also tends to converge to a local minima [20, 21]. A feasible way for solving this problem is to choose several sets of initial values, applying the EM algorithm and finally choosing the best outcome-set as the best estimation. By doing so, it can alleviate the influence of the initial values on the algorithm but increasing the computational cost. Additionally, the EM algorithm fails to converge if one or some variances of the Gaussian mixture approach to zero as it has been demonstrated when big objects with uniform intensities have undergone segmentation [14].

On the other hand, Gradient-based methods are also computationally expensive and may easily get stuck within local minima [14]. Redner and Walkerin argued in [18]—a widely-cited article, that Newtonian methods (such as Levenberg-Marquardt, LM) should generally be preferred over EM particularly for unconstrained optimization problems. However, they must be modified in order to be used within Gaussian mixtures, where there are probabilistic constraints on the parameters [22] that may result in singularities. In the parameter space of mixture models, the singularities occur when two or more components are exactly overlapped and they can be dismissed as a single component. Recent works in [19, 23] have shown that singularities cause slow convergence in Newtonian and quasi-Newtonian methods while they are applied to determine parameters of Gaussian mixtures.

Despite gradient-based methods and the EM algorithm seem to have different mechanisms for parameter updating. Xu and Jordan [22] have established a relationship between the gradient of the log likelihood and the updating step within the parameter space while using the EM algorithm. They found that the EM algorithm can be viewed as a variable metric gradient algorithm with a projection matrix changing at each step and behaving just like a function of the current parameter value. In the EM algorithm, the new value of the parameter k + 1 is thus chosen close to the previous value k mimicking gradients methods. Therefore, the updating rule may get the EM algorithm easily stuck within local minima [23].

In this chapter, an alternative approach using an optimization algorithm based on learning automata for determining the parameters of a Gaussian mixture is presented. The Learning Automata (LA) [24, 25] is an adaptive decision making method that operate in unknown random environments while progressively improve their performance via a learning process. Since LA theorists study the optimization under uncertainty, it is very useful for optimization of multi-modal functions when the function is unknown and only noise-corrupted evaluations are available. In such algorithms, a probability density function, which is defined over the parameter space, is used to select the next point. The reinforcement signal (objective function) and the learning algorithm are used by the learning automata (LA) to update the probability density function at each stage. LA has been successfully applied to solve different sorts of engineering problems such as pattern recognition [26], adaptive control [27] signal processing [28] and power systems [29].

One main advantage of the LA method is that it does not need knowledge of the environment or any other analytical reference to the function to be optimized. Additionally, one interesting advantage of LA lies on the fact that it offers fast convergence mainly when it is considered for the estimation of many parameters [30]. Other methods such as Gradient and the EM which make use of iterative updating procedures within the parameter space, may exhibit slow convergence or local minima trapping. However LA is focused on the probability space [31] leading to global optimization [32] by allowing any element of the action set (or parameter) to be chosen. This fact actually makes LA insensitive to initial values.

Recently, more effective LA-based algorithms have been proposed for multi-modal complex function optimization [28, 30–32]. It has also been experimentally shown that the performance of such optimization algorithms is comparable to or better than the genetic algorithm (GA) in [31]. On the other hand, the algorithm known as continuous action reinforcement learning automata (CARLA) [33], has been used for parameter identification of particularly complex systems, showing the effectiveness of the approach with interesting results on adaptive control [33–36] and digital filter design [28].

In this chapter, the segmentation process is considered as an optimization problem approximating the 1-D histogram of a given image by means of a Gaussian mixture model. The operation parameters are calculated through the CARLA algorithm. Each Gaussian contained within the histogram represents a pixel class and therefore belongs to the thresholding points. The experimental results, presented in this work, demonstrate that LA exhibits fast convergence, relative low computational cost and no sensitivity to initial conditions by keeping an acceptable segmentation of the image, i.e. a better mixture approximation in comparison to the EM or gradient based algorithms.

The chapter is organized as follows. Section 6.2 presents the Gaussian approximation to the histogram while Sect. 6.3 introduces the LA algorithm. Section 6.4 shows the most important implementation issues. Experimental results for the proposed approach are presented in Sect. 6.5 and some relevant conclusions are discussed in Sect. 6.6

6.2   Gaussian Approximation

Let consider an image holding L gray levels $[0, \ldots, L-1]$ whose distribution is displayed within a histogram $h(g)$. In order to simplify the description, the histogram is normalized just as a probability distribution function, yielding:

$$h(g) = \frac{n_g}{N}, \quad h(g) > 0,$$

$$N = \sum_{g=0}^{L-1} n_g, \quad \text{and} \quad \sum_{g=0}^{L-1} h(g) = 1, \tag{6.1}$$

where n_g denotes the number of pixels with gray level g and N being the total number of pixels in the image. The histogram function can thus be contained into a mix of Gaussian probability functions of the form:

$$p(x) = \sum_{i=1}^{K} P_i \cdot p_i(x) = \sum_{i=1}^{K} \frac{P_i}{\sqrt{2\pi}\sigma_i} \exp\left[\frac{-(x - \mu_i)^2}{2\sigma_i^2}\right] \tag{6.2}$$

with P_i being the probability of class i, $p_i(x)$ being the probability distribution function of gray-level random variable x in class i, with μ_i and σ_i being the mean and standard deviation of the ith probability distribution function and K being the number of classes within the image. In addition, the constraint $\sum_{i=1}^{K} P_i = 1$ must be satisfied.

The mean square error is used to estimate the $3K$ parameters P_i, μ_i and σ_i, $i = 1, \ldots, K$. For instance, the mean square error between the Gaussian mixture $p(x_i)$ and the experimental histogram function $h(x_i)$ is now defined as follows:

$$J = \frac{1}{n} \sum_{j=1}^{n} \left[p(x_j) - h(x_j) \right]^2 + \omega \cdot \left| \left(\sum_{i=1}^{K} P_i \right) - 1 \right| \qquad (6.3)$$

Assuming an n-point histogram as in [37] and ω being the penalty associated with the constrain $\sum_{i=1}^{K} P_i = 1$.

In general, the estimation of the parameters that minimize the square error produced by the Gaussian mixture is not a simple problem. A straightforward method is to consider the partial derivatives of the error function to zero, obtaining a set of simultaneous transcendental equations [37]. However, an analytical solution is not available considering the non-linear nature of the equations. The algorithms therefore make use of an iterative approach which is based on the gradient information or maximum likelihood estimation, just like the EM algorithm. Unfortunately, such methods may also get easily stuck within local minima.

For the EM algorithm and the gradient-based methods, the new parameter point lies within a neighbourhood distance of the previous parameter point. However, this is not the case for the LA's adaptation algorithm which is based on stochastic principles. The new operating point is thus determined by a parameter probability function and therefore it can be far from the previous operating point. This gives the algorithm a higher ability to locate and pursue a global minimum.

It has been shown by many works in the literature that intelligent approaches may actually provide a satisfactory performance for image processing problems [11, 12, 38–40]. The LA approach was chosen aiming into find appropriate parameters and their corresponding threshold values, yet relying on the LA convergence characteristics and its immunity to initial values.

6.3 Learning Automata (LA)

LA operates by selecting actions via a stochastic process. Such actions operate within an environment while being assessed according to a measure of the system performance. Figure 6.1a shows the typical learning system architecture. The automaton selects an action (X) probabilistically. Such actions are applied to the environment and the performance evaluation function provides a reinforcement signal β. This is used to update the automaton's internal probability distribution whereby actions that achieve desirable performance are reinforced via an increased probability. Likewise, those underperforming actions are penalised or left unchanged depending on the particular learning rule which has been employed. Over time, the average performance of the system will improve until a given limit is reached. In terms of optimization problems, the action with the highest probability would correspond to the global minimum as demonstrated by rigorous proofs of convergence available in [24, 25].

Fig. 6.1 a Reinforcement
learning system and
b interconnected automata

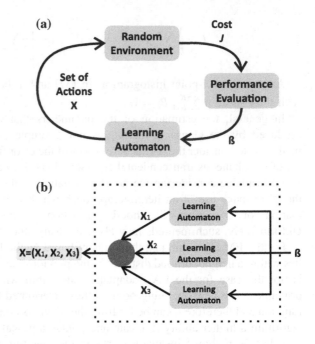

A wide variety of learning rules have been reported in the literature. One of the most widely used algorithms is the linear reward/inaction (L_{RI}) scheme, which has been shown to guaranteed convergence properties (see [24, 25]). In response to action $\mathbf{x}_i$, which is selected at time step k, the probabilities are updated as follows:

$$p_i(n+1) = p_i(n) + \theta \cdot \beta(n) \cdot (1 - p_i(n))$$
$$p_j(n+1) = p_j(n) - \theta \cdot \beta(n) \cdot p_j(n), \quad \text{if } i \neq j \tag{6.4}$$

being θ a learning rate parameter and $0 < \theta < 1$ and $\beta \in [0, 1]$ the reinforcement signal; $\beta = 1$ indicates the maximum reward and $\beta = 0$ is a null reward. Eventually, the probability of successful actions will increase to become close to unity. In case that a single and foremost successful action prevails, the automaton is deemed to have converged.

With a large number of discrete actions, the probability of selecting any particular action becomes low and the convergence time can become excessive. To avoid this, LA can be connected in a parallel setup as the one shown in Fig. 6.1b. Each automaton operates a smaller number of actions and the 'team' works together in a co-operative manner. This scheme can also be used where multiple actions are required.

Discrete stochastic LA can be used to determine global optimal parameters for optimization applications within multi-modal mean-square error surfaces. However, the discrete nature of the automata requires the discretization of a continuous parameter space while the quantization level tends to reduce the convergence rate.

Therefore, a sequential approach is adopted for the CARLA implementation [33], overcoming the problem by means of an initial coarse quantization. The method may be refined again by using a re-quantization around the most successful action later on.

6.3.1 CARLA Algorithm

The continuous action reinforcement learning automata (CARLA) is developed as an extension of the discrete stochastic LA for applications involving searching of continuous action space in a random environment [28]. Several CARLA can be connected in parallel similarly to discrete automata (Fig. 6.1b), in order to search multidimensional action spaces. Although the interconnection of the automata is through the environment, no direct inter-automata communication exists.

The automaton's discrete probability distribution is replaced by a continuous probability density function which is used as the basis for action selection. It operates a reward/inaction learning rule similar to the discrete LA shown in Eq. (6.4). Successful actions receive an increase on the probability of being selected in the future via a Gaussian neighborhood function which augments the probability density in the vicinity of such successful action. Table 6.1 shows the generic pseudo-code for the CARLA algorithm. The initial probability distribution can be chosen as being uniform over a desired range. After a considerable number of iterations, it converges to a probability distribution with a global maximum around the best action value [32].

If action x (parameter) is defined over the range (x_{min}, x_{max}), the probability density function $f(x, n)$ at iteration n is updated according to the following rule:

$$f(x, n+1) = \begin{cases} \alpha \cdot [f(x,n) + \beta(n) \cdot H(x,r)] & \text{if } x \in (x_{min}, x_{min}) \\ 0 & \text{otherwise} \end{cases} \quad (6.5)$$

Table 6.1 Generic pseudo-code for the CARLA algorithm

CARLA algorithm
Initialize the probability density function to a uniform distribution
Repeat
Select an action using its probability density function
Execute action on the environment
Receive cost/reward for previous action
Update performance evaluation function β
Update probability density function
Until stopping condition is reached

with α being chosen to re-normalize the distribution according to the following condition

$$\int_{x_{\min}}^{x_{\max}} f(x, n+1)dx = 1 \tag{6.6}$$

with $\beta(n)$ being again the reinforcement signal from the performance evaluation and $H(x, r)$ being a symmetric Gaussian neighborhood function centered on $r = x(n)$. It yields

$$H(x, r) = \lambda \cdot \exp\left(-\frac{(x - r)^2}{2\sigma^2}\right) \tag{6.7}$$

with λ and σ being parameters that determine the height and width of the neighborhood function. They are defined in terms of the range of actions as follows:

$$\sigma = g_w \cdot (x_{\max} - x_{\min}) \tag{6.8}$$

$$\lambda = \frac{g_h}{(x_{\max} - x_{\min})} \tag{6.9}$$

The speed and resolution of learning are thus controlled by free parameters g_w and g_h. Let action $x(n)$ be applied to the environment at iteration n, returning a cost or performance index $J(n)$. Current and previous costs are stored as a reference set $R(n)$. The median and minimum values J_{med} and $J_{\min}$ may thus be calculated by means of $\beta(n)$, which is defined as follows:

$$\beta(n) = \max\left\{0, \frac{J_{\text{med}} - J(n)}{J_{\text{med}} - J_{\min}}\right\} \tag{6.10}$$

To avoid problems with infinite storage requirements and to allow the system to adapt to changing environments, only the last m values of the cost functions are stored in $R(n)$. Equation (6.10) limits $\beta(n)$ to values between 0 and 1 and only returns nonzero values for those costs that are below the median value. It is easy to understand how $\beta(n)$ affects the learning process as follows: during the learning, the performance and the number of selecting actions can be wildly variable, generating extremely high computing costs. However, $\beta(n)$ is insensitive to such extremes and to high values of $J(n)$ resulting from a poor choice of actions. As the learning continues, the automaton converges towards more worthy regions of the parameter space as such actions are chosen to be evaluated more often. When more of such responses are being received, J_{med} gets reduced. Decreasing J_{med} in $\beta(n)$ effectively enables the automaton to refine its reference around better responses (previously received), and hence resulting in a better discrimination between selected actions.

In order to define an action value $x(n)$ which has been associated to a given probability density function, an uniformly distributed pseudo-random number z (n) is generated within the range of $[0,1]$. Simple interpolation is thus employed to equate this value to the cumulative distribution function:

$$\int_{x_{min}}^{x(n)} f(x, n)dx = z(n) \qquad (6.11)$$

For implementation purposes, the distribution is stored at discrete points with an equal inter-sample probability. Linear interpolation is used to determine values at intermediate positions (see full details in [28]).

6.4 Implementation

Four different pixel classes are used to segment the images. The idea is to show the effectiveness of the algorithm and its performance against other algorithms solving the same task. The implementation can easily be transferred to cases with a greater number of pixel classes.

To approach the histogram of an image by 4 Gaussian functions (one for each pixel class), it is necessary to calculate the optimum values of the 3 parameters $(P_i,\ \mu_i$ and $\sigma_i)$ for each Gaussian function (in this case, 12 values according to Eq. 6.6). This problem can be solved by optimizing Eq. (6.3), considering that function $p(x)$ gathers 4 Gaussian functions.

The parameters to be optimized are summarized in Table 6.2, with k_P^i being the parameter representing the a priori probability (P), k_σ^i holding the variance (σ) and k_μ^i representing the expected value (μ) of the Gaussian function i.

In the LA optimization, each parameter is considered like an Automaton which is able to choose actions. Such actions correspond to values assigned to the parameters by a probability distribution within the interval. All intervals considered in this work are defined as $k_P^i \in [0,0.5]$, $k_\sigma^i \in [0,128]$, and $k_\mu^i \in [0,255]$.

For this 12-dimensional problem, 12 different automatons will be created to represent parametric approach of the corresponding histogram. One of the main advantages of the LA algorithm regarding multi-dimensional problems is that the

Parameters			Gaussian
k_P^1	k_σ^1	k_μ^1	1
k_P^2	k_σ^2	k_μ^2	2
k_P^3	k_σ^3	k_μ^3	3
k_P^4	k_σ^4	k_μ^4	4

Table 6.2 Parameters to be optimized by the LA algorithm

automatons are coupled only through the environment, thus each automaton operates independently during optimization.

Thus, at each instant n, each automaton chooses an action according to their probability distribution which can be represented in a vector $A(n) = \{k_P^1, k_\sigma^1, k_\mu^1 \dots,$ $k_P^4, k_\sigma^4, k_\mu^4\}$. This vector represents a certain approach to the histogram. Then, the quality of the approach is evaluated (according to Eq. 6.3) and converted into a reinforcement signal $\beta(n)$ (through Eq. 6.10). After the reinforcement value $\beta(n)$ is defined as a product of the elected approach $A(n)$, the distribution of probability is updated for $n + 1$ of each automaton (according to Eq. 6.5). To simplify parameters in Eqs. (6.8) and (6.9), they will take the same value for the 12 automatons, such that $g_w = 0.02$ and $g_h = 0.3$. In this work, the optimization process considers a limit up to 2000 iterations.

The optimization algorithm can thus be described as follows:

i	Set iteration number $n=0$
ii	Define the action set $A(n)=\{ k_P^1, k_\sigma^1, k_\mu^1 \dots, k_P^4, k_\sigma^4, k_\mu^4 \}$ such that $k_P^i \in [0,0.5]$, $k_\sigma^i \in [0,128]$ and $k_\mu^i \in [0,255]$
iii	Define probability density functions at iteration n: $f(k_P^i, n), f(k_\sigma^i, n)$ and $f(k_\mu^i, n)$
iv	Initialize $f(k_P^i, n), f(k_\sigma^i, n)$ and $f(k_\mu^i, n)$ as a uniform distribution between the defined limits
v	Repeat while $n \leq 2000$
	(a)	Using a pseudo-random number generator for each automaton, select $z_P^i(n), z_\sigma^i(n)$ and $z_\mu^i(n)$ uniformly between 0 and 1
	(b)	Select $k_P^i \in [0,0.5]$, $k_\sigma^i \in [0,128]$ and $k_\mu^i \in [0,255]$ where the area under the probability density function is $\int_0^{k_P^i(n)} f(k_P^i, n) = z_P^i(n)$, $\int_0^{k_\sigma^i(n)} f(k_\sigma^i, n) = z_\sigma^i(n)$ and $\int_0^{k_\mu^i(n)} f(k_\mu^i, n) = z_\mu^i(n)$
	(c)	Evaluate the performance using Eq. (6.3)
	(d)	Obtain the minimum, J_{min}, and median, J_{med} of $J(n)$
	(e)	Evaluate $\beta(n)$ via Eq. (6.10)
	(f)	Update the probability density functions $f(k_P^i, n), f(k_\sigma^i, n)$ and $f(k_\mu^i, n)$ using Eq. (6.5)
	(g)	Increment iteration number n

The learning system searches within the 12-dimensional parameter space aiming for reducing the values of J in Eq. (6.3).

The final step is to determine the optimal threshold values T_i. In this case, the pixel classification corresponds to the maximum likelihood (ML) estimator. The classes can be determined by simple thresholding following standard methods, just as it is illustrated in Fig. 6.2.

Fig. 6.2 Thresholding points determination

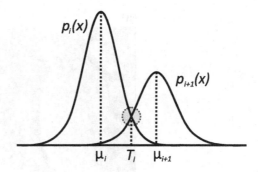

6.5 Experimental Results

This section presents the experimental work with the LA algorithm. The discussion is divided into two parts: the first one shows the performance of the proposed LA algorithm while the second discusses on a comparison between the LA segmentator, the EM algorithm and the Levenberg-Marquardt method.

6.5.1 LA Algorithm Performance in Image Segmentation

This section presents two experiments to analyze the LA's performance considering a segmentation mixture of four classes while the original histogram of the image is approached by the LA method. In order to test consistency, 10 independent repetitions are made for each experiment.

The first test considers the histogram shown by Fig. 6.3b while Fig. 6.3a presents the original image. After applying the LA algorithm (as it is explained in the previous section), a minimum is obtained (Eq. 6.3), as the point is defined by $k_P^1 = 0.094$, $k_\sigma^1 = 6$, $k_\mu^1 = 15$, $k_P^2 = 0.1816$, $k_\sigma^2 = 29$, $k_\mu^2 = 63$, $k_P^3 = 0.2733$, $k_\sigma^3 = 10$, $k_\mu^3 = 93$, $k_P^4 = 0.4503$, $k_\sigma^4 = 30$, and $k_\mu^4 = 163$. The values of such parameters define four different Gaussian functions which are clearly visible in Fig. 6.4. The original histogram and its approximation by the Gaussian mixture are visually compared in Fig. 6.5.

The evolution of the probability density parameters which in turn represent the expected values $f(k_\mu^1, n), f(k_\mu^2, n), f(k_\mu^3, n)$ and $f(k_\mu^4, n)$ of the Gaussian functions are shown in Fig. 6.6. It can be seen that most of the convergence is achieved at the first 1050 iterations, as subsequent steps yield a bit of sharpening in the distribution's shape. The final highest probability value obtained from the distribution ($n = 2000$) corresponds to the final parameter value.

From the Gaussian functions obtained by LA in Fig. 6.4, the threshold values T_i are calculated using well-known methods. Considering such values, the segmented image is shown in Fig. 6.7.

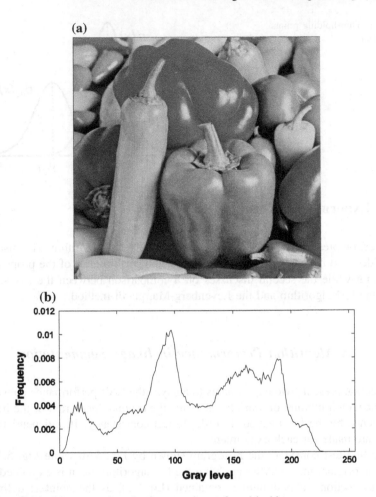

Fig. 6.3 **a** Original image used on the first experiment, **b** and its histogram

For the second experiment the image shown in Fig. 6.8 is tested. The method aims to segment the image into four different classes using the LA approach. After executing the algorithm according to the parameters defined in Sect. 6.4, the resulting Gaussian functions approximating the histogram are shown in Fig. 6.9a.

The comparison between the original image and its Gaussian approximation is shown in Fig. 6.9b. It is clear that the algorithm approaches each of all the pixel concentrations distributed within the histogram but the first one, which is presented approximately around the intensity value seven. This effect shows that the algorithm discards the smallest pixel accumulation as it prefers to cover classes that contribute to generate smaller errors during optimization of Eq. (6.3). Such results can improve if five pixel classes were used instead.

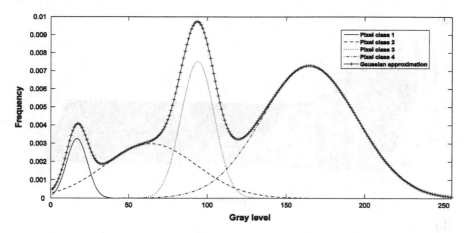

Fig. 6.4 Gaussian functions obtained by LA

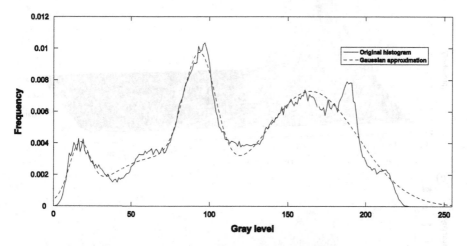

Fig. 6.5 Comparison between the original histogram and its approach

From the Gaussian mixture obtained by the LA method (Fig. 6.9a), the threshold values T_i are calculated again using well-known methods. Figure 6.10 shows the segmented image after the detection task. Figure 6.11 shows the separation of each class after applying the LA algorithm.

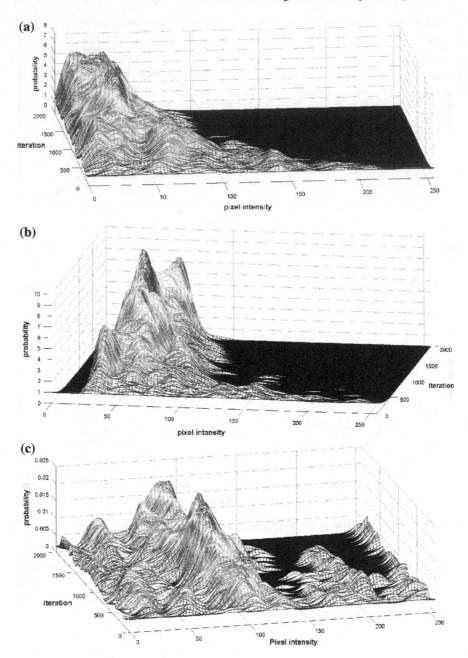

Fig. 6.6 Evolution of the probability densities parameters and their expected values of the Gaussian functions **a** $f(k_\mu^1, n)$, **b** $f(k_\mu^2, n)$, **c** $f(k_\mu^3, n)$ and **d** $f(k_\mu^4, n)$

(d)

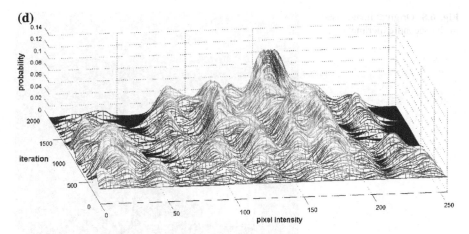

Fig. 6.6 (continued)

Fig. 6.7 Image segmented in four classes by the LA method

6.5.2 Comparing the LA Algorithm Versus the EM and LM Methods

This section discusses on the comparison between LA and other algorithms such as the EM algorithm and one Levenberg-Marquardt (LM) method. The discussion is focused on the following issues: first, sensitivity to the initial conditions; second, singularities and third, convergence and computational costs.

Fig. 6.8 Original image used in the second experiment

(a) *Sensitivity to the initial conditions.* In this experiment, initial values for all methods are initialized in different values while the same histogram is considered for the approximation task. The final parameters representing the Gaussian mixture after convergence are reported. Figure 6.12a shows the image used in this comparison while Fig. 6.12b pictures the histogram. All experiments are conducted several times in order to assure consistency. Only two different initial states with the highest variation are reported in Table 6.3. Likewise, Fig. 6.13 shows the obtained segmented images considering the two initial conditions reported by Table 6.3. In the LA case, the algorithm does not require initialization as it works with random initial values; however in order to assure a valid comparison, the same initial values are considered for the EM, the LM and the LA method.

By analyzing the information in Table 6.3, the sensitivity of the EM algorithm to initial conditions becomes evident. Figure 6.13 shows a clear pixel misclassification in some sections of the image as a consequence of such sensitivity.

(b) *Singularities.* The experiment aims to test the LA performance under certain circumstances on which it is well-reported in the literature [14, 19] that the EM and the LM have underperformed. Two cases are relevant to such purpose. First, the Gaussian variance is small or near to zero, i.e. big objects are present in the image with a homogeneous intensity value [14]. Second, the LM algorithm exhibits a slow convergence when the Gaussians are overlapped [19, 23].

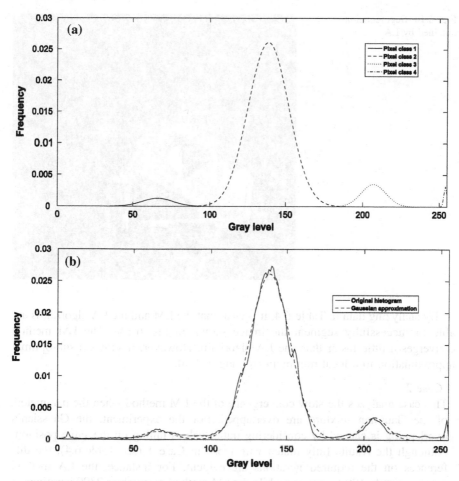

Fig. 6.9 **a** Gaussian functions obtained by the LA algorithm and **b** its comparison to the original histogram

For both cases, the EM method never reaches convergence. The benchmark image and its histogram are shown in Fig. 6.14.

Case 1.

The experiment shows the lack of convergence of the EM algorithm when a small or near to zero Gaussian variance is considered. The test consists on using all the algorithms to obtain the Gaussian mixture parameters that approximate the histogram shown in Fig. 6.14b. It is evident that only 4 classes are relevant. In order to assure consistency, the experiment is repeated over 100 times with different initial conditions. The results show that the EM method never converge to an acceptable value whatsoever. Table 6.4 shows the results for the LM and the LA algorithm as they are averaged over 100 experiments.

Fig. 6.10 Segmentation
obtained by LA

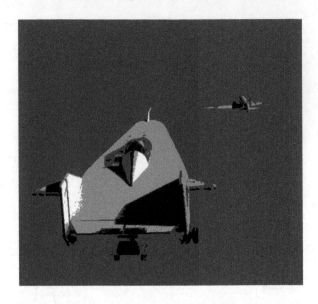

By analyzing data in Table 6.4, it is clear that the LM and the LA algorithms are able to successfully segment the image shown in Fig. 6.14a. The LM method converges a little faster than the LA algorithm. However, it shows a sub-optimal approximation to a local minimum (see Fig. 6.15a).

Case 2.
This case analyzes the slow convergence of the LM method when the parameters of the Gaussian mixture are overlapped. For the experiment, the Gaussian's overlapping is caused by considering initial values falling on the same position. Although the results fully match with those in Case 1 (see Table 6.4), the differences on the required iterations are evident. For instance, the LA method requires nearly 1000 iterations while the LM method as much as 2300 iterations—averaging 100 experiments for both cases. The convergence speed in the LA method is clearly not affected by such singularity.

(c) *Convergence and computational cost.* The experiment aims to measure the number of required steps and the computing time spent by the EM, the LM and the LA algorithm required to calculate the parameters of the Gaussian mixture in benchmark images (see Fig. 6.16a–c). All experiments consider four classes. Table 6.5 shows the averaged measurements as they are obtained from 20 experiments. It is evident that the EM is the slowest to converge (iterations) and the LM shows the highest computational cost (time elapsed) because it requires complex Hessian approximations. On the other hand, the LA shows an acceptable compromise between its convergence time and its computational cost. Finally, Fig. 6.16 shows the segmented images as they are generated by each algorithm.

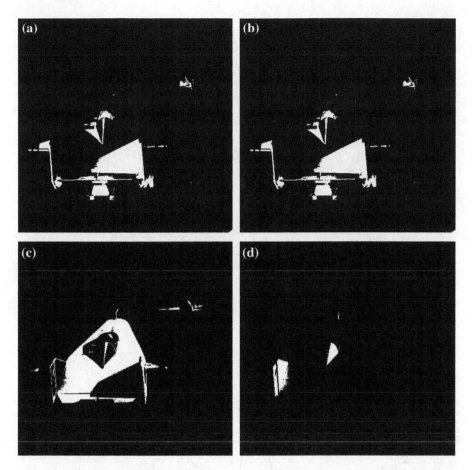

Fig. 6.11 Class separation as it is produced by the LA algorithm. **a** Pixel class 1, **b** pixel class 2, **c** pixel class 3, and **d** pixel class 4

6.6 Conclusions

In this chapter, an automatic image multi-threshold approach based on Learning Automata (LA) is proposed. The segmentation process is considered to be similar to an optimization problem. The algorithm approximates the 1-D histogram of a given image using a Gaussian mixture model whose parameters are calculated through the LA algorithm CARLA. Each Gaussian function approximating the histogram represents a pixel class and therefore one threshold point.

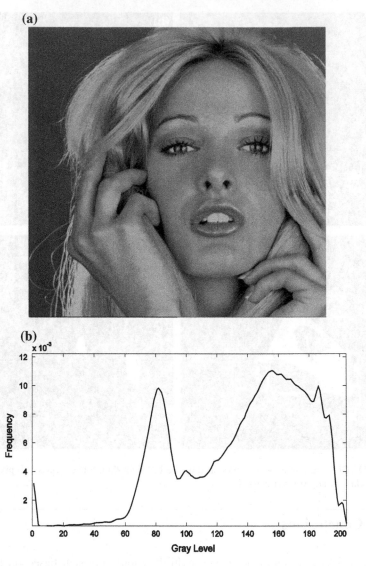

Fig. 6.12 **a** Original image used for the comparison on initial conditions and **b** its corresponding histogram

Table 6.3 Comparison between the EM, the LM and the LA algorithm, considering two different initial conditions

Parameters	Initial condition 1	EM	LM	LA	Initial condition 2	EM	LM	LA
k_μ^1	40.6	33.13	32.12	32.10	10	20.90	31.80	32.92
k_μ^2	81.2	81.02	82.05	82.01	100	82.78	80.85	82.12
k_μ^3	121.8	127.52	127	126.95	138	146.67	128	127.01
k_μ^4	162.4	167.58	166.80	166.72	200	180.72	165.90	166.62
k_σ^1	15	25.90	25.50	25.51	10	18.52	20.10	25.11
k_σ^2	15	9.78	9.70	9.65	5	12.52	9.81	9.68
k_σ^3	15	17.72	17.05	17.10	8	20.5	15.15	17.12
k_σ^4	15	17.03	17.52	17.55	22	10.09	18.00	17.15
k_P^1	0.25	0.0313	0.0310	0.312	0.20	0.0225	0.0312	0.312
k_P^2	0.25	0.2078	0.2081	0.2078	0.30	0.2446	0.2079	0.2088
k_P^3	0.25	0.2508	0.2500	0.2510	0.20	0.5232	0.2502	0.2500
k_P^4	0.25	0.5102	0.5110	0.5103	0.30	0.2098	0.5108	0.5103

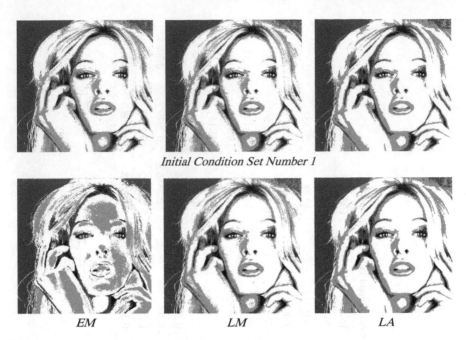

Initial Condition Set Number 1

        EM   *LM*   *LA*

Fig. 6.13 Segmented images after applying the EM, the LM and the LA algorithm with different initial conditions

Experimental evidence shows that LA algorithm has an acceptable compromise between its convergence time and its computational cost when it is compared to the Expectation-Maximization (EM) method and the Levenberg-Marquardt (LM) algorithm. Additionally, the LA algorithm also exhibits a better performance under certain circumstances (singularities) on which it is well-reported in the literature [14, 19] that the EM and the LM have underperformed. Two cases are reported: First, when Gaussian variance is small or near to zero (i.e. big objects are presented on the image with a homogeneous intensity value). Second, it is when the parameters of the Gaussian mixture are overlapped. Finally, the results have shown that the stochastic search accomplished by the LA method shows a consistent performance with no regard of the initial value and still showing a greater chance to reach the global minimum.

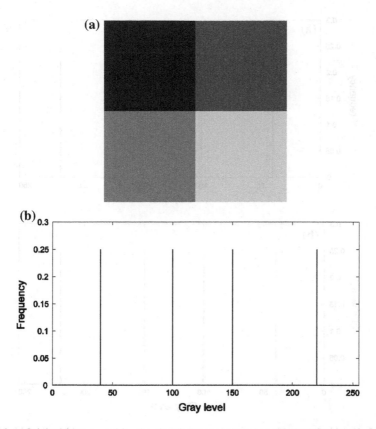

Fig. 6.14 **a** Original image used by the singularity experiment, and **b** its histogram

Table 6.4 Comparison between the LM and the LA algorithms using variances values close to zero

Parameters	LM	LA
$\bar{k}_\mu^1$	42.6	40.1
$\bar{k}_\mu^2$	98.3	99.89
$\bar{k}_\mu^3$	153.7	150.05
$\bar{k}_\mu^4$	220.1	220.01
$\bar{k}_\sigma^1$	7	0.05
$\bar{k}_\sigma^2$	12	0.07
$\bar{k}_\sigma^3$	5	0.10
$\bar{k}_\sigma^4$	0.3	0.03
$\bar{k}_P^1$	0.20	0.0313
$\bar{k}_P^2$	0.3	0.2078
$\bar{k}_P^3$	0.25	0.2508
$\bar{k}_P^4$	0.25	0.5102
Iterations	997	1050

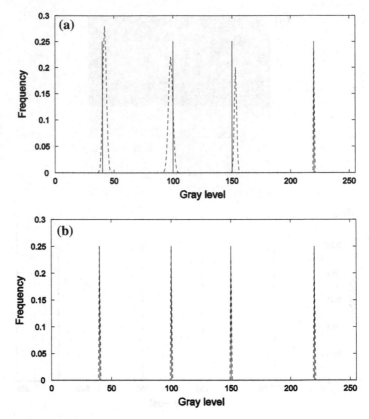

Fig. 6.15 Graphical view of approximations using near zero variances with: **a** the LM algorithm and **b** the LA method

Fig. 6.16 Original benchmark images (**a–d**), and segmented images obtained by the EM, the LM and the LA algorithms

Table 6.5 Iterations and time requirements of the EM, the LM and the LA algorithm as they are applied to segment benchmark images (see Fig. 6.16)

Iterations	(a)	(b)	(c)	(d)
Time elapsed				
EM	1855	1833	1861	1870
	2.72 s	2.70 s	2.73 s	2.73 s
LM	985	988	945	958
	4.03 s	4.04 s	4.98 s	4.98 s
LA	970	991	951	951
	1.51 s	1.53 s	1.48 s	1.48 s

References

1. Abak T, Baris U, Sankur B (1997) The performance of thresholding algorithms for optical character recognition. In: Proceedings of international conference on document analytical recognition, pp 697–700
2. Kamel M, Zhao A (1993) Extraction of binary character/graphics images from grayscale document images, Graph. Models Image Process 55(3):203–217
3. Trier OD, Jain AK (1995) Goal-directed evaluation of binarization methods. IEEE Trans Pattern Anal Mach Intel 17(12):1191–1201
4. Bhanu B (1986) Automatic target recognition: state of the art survey. IEEE Trans Aerosp Electron Syst 22:364–379
5. Sezgin M, Sankur B (2001) Comparison of thresholding methods for non-destructive testing applications. In: IEEE international conference on image processing, pp 764–767
6. Sezgin M, Tasaltin R (2000) A new dichotomization technique to multilevel thresholding devoted to inspection applications. Pattern Recogn Lett 21(2):151–161
7. Guo R, Pandit SM (1998) Automatic threshold selection based on histogram modes and discriminant criterion. Mach Vis Appl 10:331–338
8. Pal NR, Pal SK (1993) A review on image segmentation techniques. Pattern Recogn 26:1277–1294
9. Shaoo PK, Soltani S, Wong AKC, Chen YC (1988) Survey: a survey of thresholding techniques. Comput Vis Graph Image Process 41:233–260
10. Snyder W, Bilbro G, Logenthiran A, Rajala S (1990) Optimal thresholding: a new approach. Pattern Recogn Lett 11:803–810
11. Chen S, Wang M (2005) Seeking multi-thresholds directly from support vectors for image segmentation. Neurocomputing 67(4):335–344
12. Chih-Chih L (2006) A novel image segmentation approach based on particle swarm optimization. IEICE Trans Fundam 89(1):324–327
13. Böhning D, Seidel W (2003) Recent developments in mixture models. Comput Stat Data Anal 41:349–357
14. Gupta L, Sortrakul T (1998) A Gaussian-mixture-based image segmentation algorithm. Pattern Recogn 31(3):315–325
15. Dempster AP, Laird NM, Rubin DB (1977) Maximum likelihood from incomplete data via the EM algorithm. J R Stat Soc Ser B 39(1):1–38
16. Zhang Z, Chen C, Sun J, Chan L (2003) EM algorithms for Gaussian mixtures with split-and-merge operation. Pattern Recogn 36:1973–1983
17. Park H, Amari S, Fukumizu K (2000) Adaptive natural gradient learning algorithms for various stochastic models. Neural Netw 13:755–764

18. Redner RA, Walker HF (1984) Mixture densities, maximum likelihood and the EM algorithm. SIAM Rev 26(2):195–239
19. Park H, Ozeki T (2009) Singularity and slow convergence of the EM algorithm for Gaussian mixtures. Neural Process Lett 29:45–59
20. Ma J, Xu L, Jordan MI (2000) Asymptotic convergence rate of the EM algorithm for Gaussian mixtures. Neural Comput 12:2881–2907
21. Xu L, Jordan MI (1996) On convergence properties for the EM algorithm. Neural Comput 8:129–151
22. Xu L, Jordan MI (1996) On corvengence of the EM algorithm for Gaussian mixtures. Neural Comput 8(1):129–151
23. Olsson R, Petersen K, Lehn-Schiøler T (2008) State-Space models: from the EM algorithm to a gradient approach. Neural Comput 19(4):1097–1111
24. Narendra KS, Thathachar MAL (1989) Learning automata: an introduction. Prentice-Hall, London
25. Najim K, Poznyak AS (1994) Learning automata—theory and applications. Pergamon Press, Oxford
26. Seyed-Hamid Z (2008) Learning automata based classifier. Pattern Recogn Lett 29:40–48
27. Zeng X, Zhou J, Vasseur C (2000) A strategy for controlling non-linear systems using a learning automaton. Automatica 36:1517–1524
28. Howell M, Gordon T (2001) Continuous action reinforcement learning automata and their application to adaptive digital filter design. Eng Appl Artif Intell 14:549–561
29. Wu QH (1995) Learning coordinated control of power systems using inter-connected learning automata. Int J Electr Power Energy Syst 17:91–99
30. Thathachar MAL, Sastry PS (2002) Varieties of learning automata: an overview. IEEE Trans Syst Man Cybern Part B: Cybern 32:711–722
31. Zeng X, Liu Z (2005) A learning automaton based algorithm for optimization of continuous complex function. Inf Sci 174:165–175
32. Beygi H, Meybodi MR (2006) A new action-set learning automaton for function optimization. Int J Franklin Inst 343:27–47
33. Frost GP (1998) Stochastic optimization of vehicle suspension control systems via learning automata. PhD thesis, Department of Aeronautical and Automotive Engineering, Loughborough University, Loughborough, Leicestershire, LE81 3TU, UI, October 1998
34. Howell MN, Frost GP, Gordon TJ, Wu QH (1997) Continuous action reinforcement learning applied to vehicle suspension control. Mechatronics 7(3):263–276
35. Howell MN, Best MC (2000) On-line PID tuning for engine idle-speed control using continuous action reinforcement learning automata. Control Eng Pract 8:147–154
36. Kashki M, Abdel-Magid Y, Abido M (2008) A reinforcement learning automata optimization approach for optimum tuning of PID controller in AVR system. In: Huang D-S et al (eds) Advanced intelligent computing theories and applications. With aspects of artificial intelligence. ICIC 2008, LNAI 5227, pp 684–692
37. Gonzalez RC, Woods RE (1992) Digital image processing. Addison Wesley, Reading
38. Baştürk A, Günay E (2009) Efficient edge detection in digital images using a cellular neural network optimized by differential evolution algorithm. Expert Syst Appl 36(8):2645–2650
39. Lai C-C, Tseng D-C (2001) An optimal L-filter for reducing blocking artifacts using genetic algorithms. Signal Process 81(7):1525–1535
40. Tseng D-C, Lai C-C (1999) A genetic algorithm for MRF-based segmentation of multispectral textured images. Pattern Recogn Lett 20(14):1499–1510

Chapter 7
Real-Time Gaze Control Using Neurofuzzy Prediction System

Real-time gaze control is a complicated task because it considers different dynamics behaviors of the elements involved in the process. This chapter describes the use of the adaptive network-based fuzzy inference system (ANFIS) model to reduce the delay effects in gaze control. The approach considers the modelling of the target objet location to predict the future positions, in order to diminish its delay consequences. The approach has been tested in a vision system of a humanoid robot. The predictions presented in the experimental results show that the object tracking performance is better in terms of velocity and accuracy.

7.1 Introduction

Real-time gaze control is a complicated task because of the different dynamics of the elements involved in the process. On the one hand, the algorithms for image processing are usually very time-consuming. On the other hand, the motors and mechanism used to control camera movements are very slow. A gaze control system is composed of three connected subsystems. First, it has an algorithm which receives the captured image from a camera as its input. This algorithm segments the image and locates the object of interest. The localization of the object can be considered as the output of this block. The next block is the controller, which takes the object localization as its input and tries to maintain the object within the visual frame by sending the appropriate signals to the third subsystem, namely the mechanisms which can directly manipulate the camera position. This system can therefore be considered as a feedback control system in which the elements that make up the system have different dynamic characteristics. Figure 7.1 shows a representation of a gaze control system.

The development of the gaze control system poses a challenge in terms of the controller design: the controller should be robust and immune to noise due to object movement, while simultaneously working with inherent delay. The delay is in fact

© Springer International Publishing AG 2017   129
M.-A. Díaz-Cortés et al., *Engineering Applications of Soft Computing*,
Intelligent Systems Reference Library 129, DOI 10.1007/978-3-319-57813-2_7

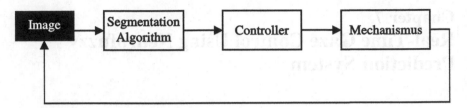

Fig. 7.1 Basic representation of gaze control system

the sum of delays produced in two system blocks, namely, image capture and image processing. These blocks are responsible for considerable delay, caused partly by time-expensive segmentation techniques and partly by the mechanisms and motors that manipulate the camera position. The magnitude of the latter delay depends on the characteristics of the particular device. A common approach to solving the delay problem is to restrict the segmentation operation to relatively simple algorithms, while using motors and expensive hardware for image capture to ensure a better dynamic behaviour. However, this approach could limit the application possibilities of the gaze control systems.

The concept of motor prediction was first introduced by Helmholtz in an attempt to understand how humans localize visual objects [1, 2]. His suggestion was that the brain predicts the gaze position of the eye, rather than sensing it. In his model, the predictions are based on a copy of the motor commands acting on the eye muscles. In effect, the gaze position of the eye is made available before sensory signals become available. A simple approach for implementing the predictor for handling linear effects. However, it is inappropriate for systems that contain significant nonlinear effects. This chapter presents a neurofuzzy prediction algorithm to eliminate the delay problem. This algorithm is able to predict the dynamics of a target object in up to six video frames. This timeframe is sufficient for most applications and could be improved without a great additional effort. The system has been tested in the vision system of a humanoid robot presenting successful results. The predictions improve the velocity and accuracy of object tracking.

The chapter is organized as follows: In Sect. 7.2 is described the neurofuzzy model while Sect. 7.3 explains the whole system implementation. Section 7.4 presents the obtained results and conclusions are presented in Sect. 7.5.

7.2   Adaptive Neurofuzzy Inference System

In this section, a class of adaptive networks [3] that are functionally equivalent to fuzzy inference is described. The proposed architecture is referred to as ANFIS [4], which stands for adaptive network-based fuzzy inference system. A description of how to decompose the parameter set to facilitate the hybrid learning rule for the ANFIS architecture representing both the Sugeno and Tsukamoto fuzzy models is presented.

7.2.1   ANFIS Architecture

For simplicity, the fuzzy inference system under consideration is assumed to have two inputs, x and y, and one output, z. For a first-order Sugeno fuzzy model, a common rule set with two fuzzy if-then rules is as follows:

$$\text{Rule } 1\text{: If } x \text{ is } A_1 \text{ and } y \text{ is } B_1, \text{ then } f_1 = p_1 x + q_1 y + r_1$$
$$\text{Rule } 2\text{: If } x \text{ is } A_2 \text{ and } y \text{ is } B_2, \text{ then } f_2 = p_2 x + q_2 y + r_2$$

Figure 7.2 illustrates the reasoning mechanism for this Sugeno model. The corresponding equivalent ANFIS architecture is shown in Fig. 7.3, where nodes of the same layer have similar functions, as described next.

Fig. 7.2 A two-input first-order Sugeno fuzzy model with two rules

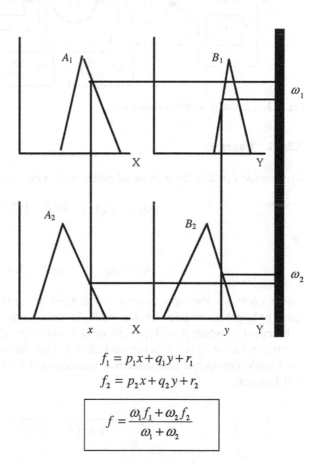

$$f_1 = p_1 x + q_1 y + r_1$$
$$f_2 = p_2 x + q_2 y + r_2$$

$$f = \frac{\omega_1 f_1 + \omega_2 f_2}{\omega_1 + \omega_2}$$

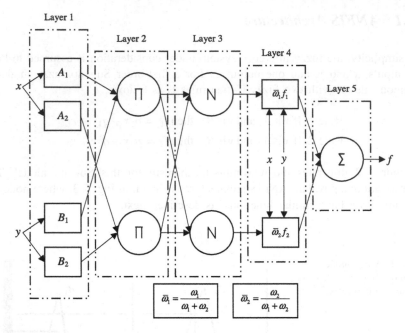

Fig. 7.3 Equivalent ANFIS architecture

7.2.1.1   Layer 1

Every node i in this layer is an adaptive node with a node function

$$O_{1,i} = \mu_{A_i}(x) \quad \text{for } i = 1, 2 \tag{7.1}$$

or

$$O_{1,i} = \mu_{B_{i-2}}(y) \quad \text{for } i = 3, 4 \tag{7.2}$$

where x (or y) is the input to node i, and A_i (or B_{i-2}) is a linguistic label (such as small or large) associated with this node. In other words, i is the membership grade of fuzzy set A (where $A = A_1, A_2, B_1$ or B_2), and it specifies the degree to which the given input x (or y) satisfies the quantifier A. Here the membership function for A can be any appropriate parameterized membership function, such as the generalized bell function

$$\mu_A(x) = \frac{1}{1 + |(x - c_i)/a_i|^{2b}} \tag{7.3}$$

where $\{a_i, b_i, c_i\}$ is the parameter set. As the values of these parameters change, the bell-shaped function varies accordingly, thus exhibiting various forms of membership functions for fuzzy set A. Parameters in this layer are referred to as premise parameters.

7.2.1.2 Layer 2

Every node in this layer is a fixed node labelled Π, whose output is the product of all the incoming signals:

$$O_{2,i} = \omega_i = \mu_{A_i}(x)\mu_{B_i}(y), \quad \text{for } i = 1, 2 \tag{7.4}$$

Each node output represents the firing strength of a rule. In general, any other T-norm operators that perform fuzzy AND can be used as the node function in this layer.

7.2.1.3 Layer 3

Every node in this layer is a fixed node labelled N, the i-th node calculates the ratio of the firing strength of the i-th rule to the sum of the firing strengths of all rules:

$$O_{3,i} = \overline{\omega}_i = \frac{\omega_i}{\omega_1 + \omega_2}, \quad \text{for } i = 1, 2 \tag{7.5}$$

For convenience, outputs of this layer are called *normalized firing strenghts*.

7.2.1.4 Layer 4

Every node i in this layer is an adaptive node with a node function:

$$O_{4,i} = \overline{\omega}_i f_i = \overline{\omega}_i(p_i x + q_i y + r_i), \tag{7.6}$$

where $\overline{\omega}_1$ is a normalized firing strength from layer 3 and $\{p_i, q_i, r_i\}$ is the parameter set of this node. Parameters in this layer are referred to as *consequent parameters*.

7.2.1.5 Layer 5

The single node in this layer is a fixed node labelled Σ, which computes the overall output as the summation of all incoming signals:

$$O_{5,i} = \sum_i \overline{\omega}_i f_i = \frac{\sum_i \omega_i f_i}{\sum_i \omega_i}. \tag{7.7}$$

Table 7.1 Learning parameters of the ANFIS algorithm

Algorithm	Forward pass	Backward pass
Premise parameters	Fixed	Gradient descent
Consequent parameters	LS estimator	Fixed
Signals	Node outputs	Error signals

7.2.2 Hybrid Learning Algorithm

From the ANFIS architecture shown in Fig. 7.3, it can be observed that the values of the premise parameters are fixed and the overall output can be expressed as a linear combination of the consequent parameters. In symbols, the output f in Fig. 7.3 can be rewritten as

$$
\begin{aligned}
f &= \frac{\omega_1}{\omega_1 + \omega_2} f_1 + \frac{\omega_2}{\omega_2 + \omega_2} f_2 \\
&= \overline{\omega_1}(p_1 x + q_1 y + r_1) + \overline{\omega_2}(p_2 x + q_2 y + r_2) \\
&= (\overline{\omega_1}x)p_1 + (\overline{\omega_1}y)q_1 + (\overline{\omega_1})r_1 + (\overline{\omega_2}x)p_2 + (\overline{\omega_2}y)q_2 + (\overline{\omega_2})r_2,
\end{aligned}
\tag{7.8}
$$

which is linear in the consequent parameters p_1, q_1, p_2, q_2, and r_2. From this observation come the following points:

- S is the set of total parameters,
- S_1 is the set of premise (nonlinear) parameters, and
- S_2 is the set of consequent (linear) parameters.

The learning algorithm for ANFIS is a hybrid algorithm that combines gradient descent and least-squares (LS) methods. More specifically, in the forward pass of the hybrid learning algorithm, node outputs go forward until layer 4, and the consequent parameters are identified by the least-squares method. In the backward pass, the error signals propagate backward, and the premise parameters are updated by gradient descent. Table 7.1 summarizes the activities in each pass.

The consequent parameters are identified as optimal under the condition that the premise parameters are fixed. Accordingly, the hybrid approach converges much faster, since it reduces the search-space dimensions of the original pure back-propagation method.

7.3 Implementation

7.3.1 Description

For this work, the gaze problem is applied to a soccer ball whose movements can be strong or violent. The objective is to maintain visual contact by keeping the object in the image frame. The mechanism which manipulates the camera position

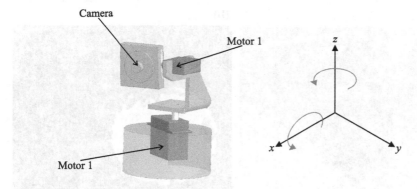

Fig. 7.4 The tracking mechanism

(Fig. 7.4) consists of two aluminum links coupled to two motors in such a way that the complete system has two degrees of freedom. The first link is coupled to motor 1, allowing angular movement along the z-axis, while the second link is coupled to motor 2 and the camera, allowing movement along the x-axis. Using this configuration to appropriately control the movement of motor 1, one can track the x-coordinate of the object within the image, while appropriate control of motor 2 can accomplish tracking of the y-coordinate. This mechanism comprises the vision system of a humanoid robot developed at the Freie Universität Berlin. In the next section, the development and content of the blocks which integrate he tracking system are explained.

7.3.2 Segmentation Algorithm and Localization

A colour segmentation algorithm is chosen because the soccer ball possesses a homogeneous colour pattern. Thus a simple change of colour model and thresholding of the image are sufficient for the segmentation of the ball. Figure 7.5a shows the object to be segmented. The image captured by the camera in red-green-blue (RGB) format (reproduced here in grayscale) is transformed to a hue-saturation-value (HSV) model. The HSV model is convenient for colour segmentation because colours that are similar are grouped together [5–7]. The three-dimensional model can be easily reduced to one dimension by dropping the saturation and value coordinates when pure and light colours are important, or by dropping the hue and value coordinates when pure and dark colours are important. Because the soccer ball has a dark colour (orange in the original colour photo), only the S plane is taken from the HSV planes. From the histogram of the image (containing the object to be considered), it is observed that an ideal threshold value is 220. Figure 7.5b shows the result of this process.

(a)　　　　　　　　　　　　　　　　(b)

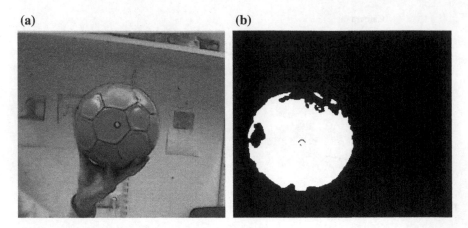

Fig. 7.5 **a** object to be segmented, **b** segmented ball

Once the object is located, the moment method can be used to identify the centoid. The equations employed are

$$M_{00} = \sum_x \sum_y I(x,y), \quad M_{10} = \sum_x \sum_y x I(x,y),$$

$$M_{01} = \sum_x \sum_y y I(x,y), \tag{7.9}$$

$$x_c = \frac{M_{10}}{M_{00}}, \quad y_c = \frac{M_{01}}{M_{00}}$$

where M_{00} represents the zero-degree moment, M_{10} and M_{01} represent the first-degree moments of x and y respectively, and x_c and y_c represent the centre coordinates.

7.3.3　Controller

A proportional-integral-derivative (PID) controller is used to control the movement of each axis (a fuzzy controller or adaptive control can also be used, as described in [8]). The objective of the gaze control is to maintain the target object inside the image frame. Thus, the controller inputs are the differences between the object localization point and the central point of the image. Figure 7.6 shows this process.

The system configuration shown in Fig. 7.6 works poorly because of the delay generated by the vision algorithms and the mechanical system. This work proposes to eliminate this problem by incorporating a neurofuzzy predictor system to predict the object dynamics. Thus, the system is modified as shown in Fig. 7.7.

Here the ANFIS system predicts the object dynamics in up to six frames, thereby compensating for the inherent delays of the system.

Fig. 7.6 Controller
configuration

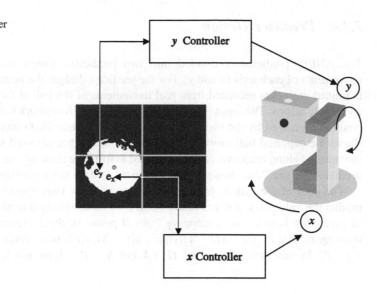

Fig. 7.7 Modified system

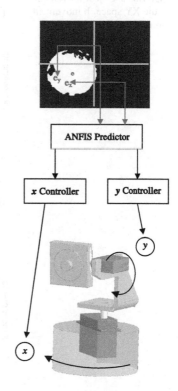

7.3.4   Predictor Design

The ANFIS predictor is divided into two predictors which correspond to the movements of each axis (x and y). For the predictor design, the neurofuzzy network is trained with data recorded from real movements of the ball at the desired speeds and accelerations. The input data presented to the ANFIS network is based on the ball position perceived by the vision module in approximately 2000 frames. Figure 7.8a shows the captured ball movements in the XY space that are used for training. The movements along each axis can be separated to form the training sets for each ANFIS predictor. Figure 7.8b shows the data set used to train the x predictor.

The goal of the task is to use past values of the x movements up to time t to predict the value of x at some future point $t + P$. The standard method for this type of prediction is to create a mapping from D points of the x movements spaced D units apart, that is $[x(t - (D - 1)\Delta), \ldots, x(t - \Delta), x(t)]$, to a predicted future value $x(t + P)$. In this work, the values $D = 4$ and $\Delta = P = 6$ are used.

Fig. 7.8 a Object movement in the XY space. **b** movement along the x-axis

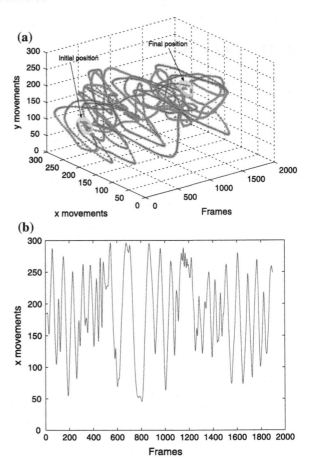

From the recorded data $x(t)$, 1000 input-output data pairs of the following format are extracted:

$$[x(t-18), x(t-12), x(t-6), x(t), x(t+6)],   (7.10)$$

where $t = 118$–1117. The first 500 data pairs are used as the training data set for ANFIS, while the remaining 500 pairs serve as the checking data set to validate the identified ANFIS. The network has four input units and one output unit, and the number of membership functions (MFs) assigned to each input of the ANFIS is two, so that the number of rules is 16. Figure 7.9 shows the learning results in the membership functions for the input $x(t-12)$. Fig. 7.9a describes the initial configuration of the membership functions, and Fig. 7.9b shows the final result of the learning process. The ANFIS used here contains a total of 104 fitting parameters, of

Fig. 7.9 Learning results in the membership functions for the input $x(t-12)$. **a** Initial configuration, **b** the final membership functions

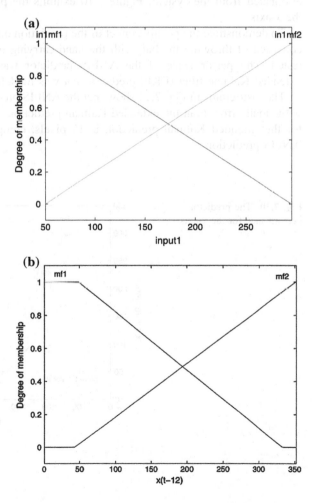

which 24 are premise (nonlinear) parameters and 80 are consequent (linear) parameters. For the ANFIS training, functions implemented with the MatLab fuzzy logic toolbox are used [9]. The learning results can be summarized as the final membership functions and the rule equations of the Sugeno inference system. All of the mentioned tasks are carried out in the same way for movements along the y-axis.

The complete system was coded in C++ and tested on a 900 MHz PCx86 with 128 MB of RAM, operating in real time on a 352×288 pixel image.

7.4 Results

The neurofuzzy predictor described in Sect. 7.3.4 was tested with high-velocity ball movements. The system performed very well, and the influence of delay was nearly eliminated from the system. Figure 7.10 exhibits the predictor's performance for the x-axis.

To demonstrate the positive effect of the prediction on the robot behavior, which consisted of throwing the ball with the hand carrying out multiple rebounds, was tested. The performance of the ANFIS predictor was compared to that of an extended Kalman filter (EKF) predictor that was modelled with the same data.

The histograms in Fig. 7.11 show that the ANFIS prediction has far more frames with small errors than the extended Kalman prediction. The average position error for the extended Kalman prediction is 15 pixels, compared to six pixels for the ANFIS prediction.

Fig. 7.10 The predictor performance for the x-axis

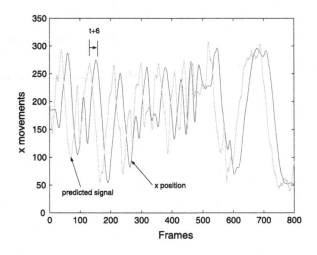

Fig. 7.11 Comparison between the histograms of the extended Kalman filter and ANFIS for predicted ball position

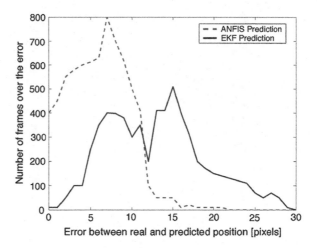

7.5 Conclusions

This chapter described the use of a predictor based in the ANFIS model to reduce the delay effects in gaze control. The presented method for motion prediction is based on a neurofuzzy system.

A neurofuzzy system for predicting the motion of a target object was successfully developed, implemented and tested. The prediction compensates for the system delay and thus allows precise and fast motion control.

Experimental results showed that this system is able to correctly predict the position and the motion of temporally coherent objects. Finally, the presented algorithm showed better precision when compared to the extended Kalman filter usually used in prediction applications.

References

1. Miall RC, Weir DJ, Wolpert DM, Stein JF (1993) Is the cerebellum a smith predictor? J Motor Behav 25(3):203–216
2. Wolpert DM, Flanagan JR (1990) Motor prediction. Curr Biol Mag 11(18):R729–R732
3. Rojas Raul (1996) Neural networks: a systematic introduction. Springer, Heidelberg
4. Jang SR, Sun CT, Mitzutani E (1998) Neurofuzzy and soft computing. Prentice Hall, New York
5. Teichner WH (1970) Color and information coding. Proc Soc Inf Displays 20:310–322
6. Watt AH (1989) Fundamentals of three-dimensional computer graphics. Addison-Wesley, Wokingham
7. Wyszecki G, Siles WS (1982) Color science, 2nd edn. Wiley, New York
8. Cuevas E, Zaldivar D, Rojas R (2004) Fuzzy condensed algorithm applied to control robotic head fr visual tracking. In: Proceedings of the IEEE international symposium robotics and automation (ISRA 2004), pp 52–56
9. The MathWorks (1999) Fuzzy logic toolbox. The MathWorks, New York. http://www.mathworks.com/access/helpdesk/help/toolbox/fuzzy/index.html?/access/helpdesk/help/toolbox/fuzzy/index.html

Chapter 8
Clonal Selection Algorithm Applied to Circle Detection

Automatic circle detection in digital images is considered an important and complex task for the computer vision community. Consequently, recently, a tremendous amount of research has been devoted to find an optimal circle detector. This chapter presents an algorithm for the automatic detection of circular shapes from complicated and noisy images without making use of the conventional Hough transform principles. The presented algorithm is based on the artificial immune optimization technique, known as the clonal selection algorithm (CSA). The CSA is an effective method for searching and optimizing following the clonal selection principle in the human immune system which generates a response according to the relationship between antigens (Ag), i.e. patterns to be recognized and antibodies (Ab) i.e. possible solutions. The algorithm uses the encoding of three points as candidate circles (x, y, r) over the edge image. An objective function evaluates if such candidate circles (Ab) are actually present in the edge image (Ag). Guided by the values of this objective function, the set of encoded candidate circles are evolved using the CSA so that they can fit to the actual circles on the edge map of the image. Experimental results over several synthetic as well as natural images with varying range of complexity validate the efficiency of the presented technique with regard to accuracy, speed, and robustness.

8.1 Introduction

Bio-inspired computing [1] lies within the realm of natural computing, a field of research that is concerned with both the use of biology as an inspiration for solving computational problems and the use of the natural world experiences to solve real world problems. The increasing interest in this field lies in the fact that nowadays the world is facing more and more complex, large, distributed and ill-structured systems, while on the other hand, people notice that the apparently simple structures and organizations in nature are capable of dealing with most complex systems and

© Springer International Publishing AG 2017   143
M.-A. Díaz-Cortés et al., *Engineering Applications of Soft Computing*,
Intelligent Systems Reference Library 129, DOI 10.1007/978-3-319-57813-2_8

tasks with ease. Bio-inspired computing has proved to be useful in various application areas. Following features from optimization, pattern recognition, shape detection and machine learning, the Bio-inspired algorithms have recently gained considerable research interest from the computer vision community. Currently, bio-inspired algorithms are widely applied to solve challenging computer vision problems. For instance, Chih-Chih [2] have applied the particle swarm optimization (PSO) algorithm for image segmentation. Le Hégarat-Mascle et al. [3] proposed a non-stationary Markov model-based image regularization algorithm, which uses another swarm intelligence algorithm known as ant colony optimization (ACO). In [4], Hammouche et al. proposed a multilevel method that allows the determination of the appropriate number of thresholds for image segmentation. Such method combines a genetic algorithm (GA) with a wavelet transform. More recently, Baştürk and Günay [5] have proposed an image edge detector based on a cellular neural network which is optimized by the differential evolution (DE) algorithm.

The problem of detecting circular features holds paramount importance for image analysis, in particular for industrial applications such as automatic inspection of products and components, aided vectorization of drawings, target detection, etc. [6]. Many methods have been developed to solve the shape-detection problem [7]. Solving the object location is normally approached from two viewpoints: deterministic techniques which include the application of Hough transform [8], geometric hashing, template or model matching techniques [9, 10]. On the other hand, stochastic techniques include random sample consensus [11], simulated annealing [12] and genetic algorithms (GA) [13].

Template and model matching techniques are the first approaches to be successfully applied to shape localization. Shape coding techniques and combination of shape properties are used to represent such objects. The main drawback of these techniques is related to the contour extraction from real images. Additionally, it is difficult for models to deal with pose invariance unless only simple objects are considered.

Commonly, circle detection in digital images is performed by means of circular Hough transform [14]. A typical Hough-based approach employs an edge detector and uses edge information to infer locations and radius values. Peak detection is then applied by averaging, filtering and histogramming the transform space. However, such approach requires a large storage space-given the 3-D cells needed to store the parameters (x, y, r), the computational complexity yielding low processing speeds.

The accuracy of the detected circle's parameters is poor, under noisy conditions [15]. The required processing time for Circular Hough Transform makes it prohibitive to be deployed in real time applications, in particular for digital images with significant width and height and a densely populated area around edge pixels. In order to overcome such a problem, other researchers have proposed new approaches based on the Hough transform (HT) such as the probabilistic HT [11, 16], the randomized HT (RHT) [17] and the fuzzy Hough transform (FHT) [18]. In [19], Lu and Tan proposed a novel approach based on RHT called iterative randomized Hough transformation (IRHT) that achieves better results on complex images and noisy environments. The algorithm iteratively applies the randomized Hough transform (RHT) to a region of interest in the image which is determined from the latest estimation of ellipse/circle parameters.

Shape recognition can also be approached using stochastic search methods such as genetic algorithms (GA). In particular, GA has recently been applied to important shape detection tasks e.g. Roth and Levine [13] proposed use of GA for primitive extraction of images. Lutton et al. developed a further improvement of the aforementioned method [20]. Yao et al. [21] came up with a multi-population GA method to detect ellipses. In [19], GA was used for template matching when the pattern has been the subject of an unknown affine transformation. Ayala–Ramirez et al. [22] presented a GA-based circle detector that is capable of detecting multiple circles on real images but it fails frequently when detecting imperfect circles. Recently, Dasgupta et al. [23] in an excellent work proposed an automatic circle detector using the bacterial foraging algorithm as optimization method. For the case of ellipsoidal detection, Rosin [24] proposed in an ellipse fitting algorithm that uses five points.

On the other hand, biological inspired methods can successfully be transferred into novel computational paradigms as shown by the successful development of artificial neural networks, evolutionary algorithms, swarming algorithms and so. The human immune system (HIS) is a highly evolved, parallel and distributed adaptive system [25] that exhibits remarkable abilities that can be imported into important aspects in the field of computation. This emerging field is known as artificial immune systems (AIS) [26] which is a computational system fully inspired by the immunology theory and its functions, including principles and models. AIS have recently reached considerable research interest from different communities [27], focusing on several aspects of optimization, pattern recognition, abnormality detection, data analysis and machine learning. Artificial immune optimization has been successfully applied to tackle numerous challenging optimization problems with remarkable performance in comparison to other classical techniques [28].

Clonal selection algorithm (CSA) [29] is one of the most widely employed AIO approaches. The CSA is a relatively novel evolutionary optimization algorithm which has been built on the basis of the clonal selection principle (CSP) [30] of HIS. The CSP explains the immune response when an antigenic pattern is recognized by a given antibody. In the clonal selection algorithm, the antigen (Ag) represents the problem to be optimized and its constraints, while the antibodies (Ab) are the candidate solutions of the problem. The antibody-antigen affinity indicates as well the matching between the solution and the problem. The algorithm performs the selection of antibodies based on affinity either by matching against an antigen pattern or by evaluating the pattern via an objective function. In mathematical grounds, CSA has the ability of getting out of local minima while simultaneously operating over a pool of points within the search space. It does not use the derivatives or any of its related information as it employs probabilistic transition rules instead of deterministic ones. Despite to its simple and straightforward implementation, it has been extensively employed in the literature for solving several kinds of challenging engineering problems [31–33].

This chapter presents an algorithm for the automatic detection of circular shapes from complicated and noisy images with no consideration of the conventional Hough transform principles. The presented algorithm is based on a recently developed artificial immune optimization (AIO) technique, known as the clonal selection algorithm (CSA). The algorithm uses the encoding of three non-collinear edge points as candidate circles (x, y, r) in the edge image of the scene. An objective function evaluates if such candidate circles (Ab) are actually present in the edge image (Ag). Guided by the values of this objective function, the set of encoded candidate circles are evolved using the CSA so that they can fit into the actual circles within the edge map of the image. The approach generates a sub-pixel circle detector which can effectively identify circles in real images despite circular objects exhibiting a significant occluded portion. Experimental evidence shows the effectiveness of such method for detecting circles under different conditions. Comparison to one state-of-the-art GA-based method [18] and a randomized Hough transform approach (IRHT) [12] on multiple images demonstrates a better performance of the presented method.

The chapter is organized as follows: Sect. 8.2 provides a brief CSA explanation. Section 8.3 formulates the approach and studies the main features of the CSA method as it is used to detect circles in images. Section 8.4 shows the experimental results of applying our method to the recognition of circles in different image conditions. Finally, Sect. 8.5 discusses several conclusions.

8.2   Clonal Selection Algorithm

In natural immune systems, only the antibodies (Abs) which are able to recognize the intrusive antigens (non-self cells) are to be selected to proliferate by cloning [25]. Therefore, the fundament of the clonal optimization method is that only capable Abs will proliferate. Particularly, the underlying principles of the CSA are borrowed from the CSP as follows:

- Maintenance of memory cells which are functionally disconnected from repertoire,
- Selection and cloning of most stimulated Abs,
- Suppression of non-stimulated cells,
- Affinity maturation and re-selection of clones showing the highest affinities, and
- Mutation rate proportional to Abs affinities.

From immunology concepts, an antigen is any substance that forces the immune system to produce antibodies against it. Regarding the CSA systems, the antigen concept refers to the pending optimization problem which focuses on circle detection. In CSA, B cells, T cells and antigen-specific lymphocytes are generally called antibodies. An antibody is a representation of a candidate solution for an antigen, e.g. the prototype circle in this work. A selective mechanism guarantees that those antibodies (solutions) that better recognize the antigen and therefore may elicit the response, are to be selected holding long life spans. Therefore such cells are to be named memory cells (**M**).

8.2.1 Definitions

In order to describe the CSA, the notation includes boldfaced capital letters indicating matrices and boldfaced small letters indicating vectors. Some relevant concepts are also revisited below:

(i) Antigen: the problem to be optimized and its constraints (circle detection).
(ii) Antibody: the candidate solutions of the problem (circle candidates).
(iii) Affinity: the objective function measurement for an antibody (circle matching).

The limited-length character string **d** is the coding of variable vector **x** as **d** = *encode*(**x**); and **x** is called the decoding of antibody **d** following **x** = *decode*(**d**).

Set **I** is called the antibody space namely **d** ∈ **I**. The antibody population space is thus defined as:

$$\mathbf{I}^m = \{\mathbf{D} : \mathbf{D} = (\mathbf{d}_1, \mathbf{d}_2, \ldots, \mathbf{d}_m), \mathbf{d}_k \in \mathbf{I}, 1 \leq k \leq m\} \qquad (8.1)$$

where the positive integer m is the size of antibody population $\mathbf{D} = \{\mathbf{d}_1, \mathbf{d}_2, \ldots, \mathbf{d}_m\}$ which is an m-dimensional group of antibody **d**, being a spot within the antibody space **I**.

8.2.2 CSA Operators

Based on [34], the CSA implements three different operators: the clonal proliferation operator (T_P^C), the affinity maturation operator (T_M^A) and the clonal selection operator (T_S^C). $\mathbf{A}(k)$ is the antibody population at time k that represents the set of antibodies **a**, such as $\mathbf{A}(k) = \{\mathbf{a}_1(k), \mathbf{a}_2(k), \ldots, \mathbf{a}_n(k)\}$. The evolution process of CSA can be described as follows:

$$\mathbf{A}(k) \xrightarrow{T_C^P} \mathbf{Y}(k) \xrightarrow{T_M^A} \mathbf{Z}(k) \cup \mathbf{A}(k) \xrightarrow{T_S^C} \mathbf{A}(k+1) \qquad (8.2)$$

8.2.2.1 Clonal Proliferation Operator (T_P^C)

Define

$$\mathbf{Y}(k) = T_P^C(\mathbf{A}(k)) = \left[T_P^C(\mathbf{a}_1(k)), T_P^C(\mathbf{a}_1(k)), \ldots, T_P^C(\mathbf{a}_n(k))\right] \qquad (8.3)$$

where $\mathbf{Y}(k) = T_P^C(\mathbf{A}(k)) = \mathbf{e}_i \cdot \mathbf{a}_i(k)$ $i = 1, 2, \ldots, n,$, and $\mathbf{e}_i$ is a q_i-dimensional identity column vector. Function round(x), gets x to the least integer bigger than x. There are various methods for calculating q_i. In this work, it is calculated as follows:

$$q_i(k) = \text{round}\left[N_c \cdot \frac{F(\mathbf{a}_i(k))}{\sum_{j=1}^n F(\mathbf{a}_j(k))}\right] i = 1, 2, \ldots, n \qquad (8.4)$$

where N_c is called the clonal size. The value of $q_i(k)$ is proportional to the value of $F(\mathbf{a}_i(k))$. After clonal proliferation, the population becomes

$$\mathbf{Y}(k) = \{\mathbf{Y}_1(k), \mathbf{Y}_2(k), \ldots, \mathbf{Y}_n(k)\} \qquad (8.5)$$

where

$$\mathbf{Y}_i(k) = \{\mathbf{y}_{ij}(k)\} = \{\mathbf{y}_{i1}(k), \mathbf{y}_{i2}(k), \ldots, \mathbf{y}_{iq_i}(k)\} \text{and}$$
$$\mathbf{y}_{ij}(k) = \mathbf{a}_1(k),\ j = 1, 2, \ldots, q_i.\ i = 1, 2, \ldots, n. \qquad (8.6)$$

8.2.2.2 Affinity Maturation Operator (T_M^A)

The affinity maturation operation is performed by hypermutation. Random changes are introduced into the antibodies just like it happens in the immune system. Such changes may lead to increase the affinity. The hypermutation is performed by the operator T_M^A which is applied to the population $\mathbf{Y}(k)$ as it is obtained by clonal proliferation $\mathbf{Z}(k) = T_M^C(\mathbf{Y}(k))$.

The mutation rate is calculated using the following equation [35]:

$$\alpha = e^{(-\rho \cdot F(ab))} \qquad (8.7)$$

being α the mutation rate, F being the objective function value of the antibody (ab) as it is normalized between [0,1] and ρ being a fixed step. In [36], it is demonstrated the importance of including the factor ρ into Eq. (8.7) to improve the algorithm performance. The way ρ modifies the shape of the mutation rate is shown by Fig. 8.1.

The number of mutations held by a clone with objective function value F, is equal to $L \cdot \alpha$, considering L as the length of the antibody -22 bits are used in this chapter. For the binary encoding, mutation operation can be done as follows: each gene within an antibody may be replaced by its opposite number (i.e. 0–1 or 1–0).

Fig. 8.1 Hypermutation rate versus fitness, considering some size steps

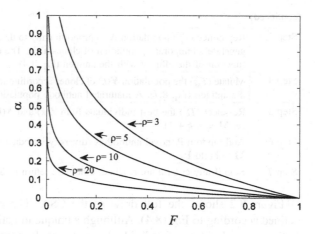

Following the affinity maturation operation, the population becomes:

$$\mathbf{Z}(k) = \{\mathbf{Z}_1(k), \mathbf{Z}_2(k), \ldots, \mathbf{Z}_n(k)\}$$
$$\mathbf{Z}_i(k) = \{\mathbf{z}_{ij}(k)\} = \{\mathbf{z}_{i1}(k), \mathbf{z}_{i2}(k), \ldots, \mathbf{z}_{iq_1}(k)\} \text{ and} \qquad (8.8)$$
$$\mathbf{z}_{ij}(k) = T_M^A(\mathbf{y}_{ij}(k)), \quad j = 1, 2, \ldots, q_1 \quad i = 1, 2, \ldots, n$$

where T_M^A is the operator as it is defined by Eq. (8.7) and applied onto the antibody $\mathbf{y}_{ij}$.

8.2.2.3 Clonal Selection Operator (T_S^C)

Define $\forall i = 1, 2, \ldots, n$, $\mathbf{b}_i(k) \in \mathbf{Z}_i(k)$ as the antibody with the highest affinity in $\mathbf{Z}_i(k)$, then $\mathbf{a}_i(k+1) = T_S^C(\mathbf{Z}_i(k) \cup \mathbf{a}_i(k))$, where T_S^C is defined as:

$$T_S^C(\mathbf{Z}_i(k) \cup \mathbf{a}_i(k)) - \begin{cases} \mathbf{b}_i(k) & \text{if } F(\mathbf{a}_i(k)) < F(\mathbf{b}_i(k)) \\ \mathbf{a}_i(k) & \text{if } F(\mathbf{a}_i(k)) \geq F(\mathbf{b}_i(k)) \end{cases} \qquad (8.9)$$

where $i = 1, 2, \ldots, n$.

Each step of the CSA may be defined as follows:

Step 1:	Initialize randomly a population (Pinit), a set $h = P_r + n$ of candidate solutions of subsets of memory cells (**M**) which is added to the remaining population (P_r), with the total population being $\mathbf{P_T} = P_r + M$, with **M** holding n memory cells
Step 2:	Select the n best individuals of the population $\mathbf{P_T}$ to build $\mathbf{A}(k)$, according to the affinity measure (objective function)

(continued)

(continued)

Step 3:	Reproduce (T_P^C) population $\mathbf{A}(k)$ proportionally to their affinity with the antigen and generate a temporary population of clones $\mathbf{Y}(k)$. The clone number is an increasing function of the affinity with the antigen (Eq. 8.4)
Step 4:	Mutate (T_M^A) the population $\mathbf{Y}(k)$ of clones according to the affinity of the antibody to the antigen (Eq. 8.7). A matured antibody population $\mathbf{Z}(k)$ is thus generated
Step 5:	Re-select (T_S^C) the best individuals from $\mathbf{Z}(k)$ and $\mathbf{A}(k)$ to compose a new memory set $\mathbf{M} = \mathbf{A}(k + 1)$
Step 6:	Add random P_r novel antibodies (diversity introduction) to the new memory cells $\mathbf{M}$ to build $\mathbf{P_T}$
Step 7:	Stop if any criteria are reached, otherwise return to Step 2

Figure 8.2 shows the full draw of the CSA. The clone number in Step 3 is defined according to Eq. (8.4). Although a unique mutation operator is used in Step 5, the mutated values of individuals are inversely proportional to their fitness by means of Eq. (8.7), i.e. the more Ab shows a better fitness, the less it may change.

The similarity property [37] within the Abs can also affect the convergence speed of the CSA. The idea of the antibody addition based on the immune network theory is introduced for providing diversity to the newly generated Abs in $\mathbf{M}$, which may be similar to those already in the old memory $\mathbf{M}$. Holding such a diverse Ab pool, the CSA can avoid being trapped into local minima [38], contrasting to well-known genetic algorithms (GA) which usually tend to bias the whole population of chromosomes towards only the best candidate solution [39]. Therefore, it can effectively handle challenging multimodal optimization tasks [40–43].

The management of population includes a simple and direct searching algorithm for globally optimal multi-modal functions. This is also another clear difference in comparison to other evolutionary algorithms, like GA, because it does not require crossover but only cloning and hyper-mutation of individuals in order to use affinity as selection mechanism. The CSA is adopted in this work in order to find the circle parameters (x, y, r) that better represent the actual circles in the image.

Fig. 8.2 Basic flow diagram of clonal selection algorithm (CSA)

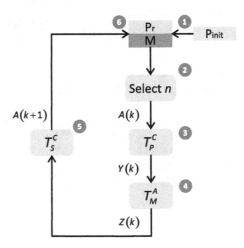

8.3 Circle Detection Using CSA

Circles are represented in this work by means of parameters of a well-known second degree equation (see Eq. 8.10), that passes through three points [13] in the edge space of the image. Images are preprocessed by an edge detection method which uses a single-pixel contour detector. Such task is accomplished by the classical Canny algorithm which stores locations for each edge point. Therefore, such points are the only potential candidates to define circles by considering triplets. All the edge points in the image are then stored within a vector array $P = \{p_1, p_2, \ldots, p_{N_p}\}$ with N_p as the total number of edge pixels contained in the image. The algorithm stores the (x_i, y_i) coordinates for each edge pixel p_i in the edge vector.

In order to construct each of the circle candidates (or antibodies within the AIS-framework), the indexes i_1, i_2 and i_3 of three non-collinear edge points must be combined, assuming the circle's contour goes through points $p_{i_1}; p_{i_2}; p_{i_3}$. A number of candidate solutions are generated randomly for the initial pool. The solutions will thus evolve through the application of the CSA as the evolution takes place over the pool until a minimum is reached and the best individual is considered as the solution for the circle detection problem.

Applying classic methods based on Hough Transform for circle detection would normally require huge amounts of memory and consume large computation time. In order to reach a sub-pixel resolution–just like the method discussed in this chapter, they also consider three edge points to cast a vote for the corresponding point within the parameter space. Such methods also require an evidence-collecting step which is also implemented by the method in this chapter. As the overall evolution process evolves, the objective function improves at each generation by discriminating non-plausible circles and locating others by avoiding a visit to other image points.

The following discussion clearly explains the required steps to formulate the circle detection task just as an AIO optimization problem.

8.3.1 Individual Representation

Each antibody C of the pool uses three edge points as elements. In this representation, the edge points are stored according to one index that is relative to their position within the edge array P. In turn, the procedure will encode an Ab as the circle that passes through three points p_i, p_j and p_k ($C = \{p_i, p_j, p_k\}$). Each circle C is represented by three parameters: x_0, y_0 and r, being (x_0, y_0) the (x, y) coordinates of the center of the circle and r its radius. The equation of the circle passing through the three edge points can thus be computed as follows:

$$(x - x_0)^2 + (y - y_0)^2 = r^2 \tag{8.10}$$

considering

$$\mathbf{A} = \begin{bmatrix} x_j^2 + y_j^2 - (x_i^2 + y_i^2) & 2 \cdot (y_j - y_i) \\ x_k^2 + y_k^2 - (x_i^2 + y_i^2) & 2 \cdot (y_k - y_i) \end{bmatrix}$$

$$\mathbf{B} = \begin{bmatrix} 2 \cdot (x_j - x_i) & x_j^2 + y_j^2 - (x_i^2 + y_i^2) \\ 2 \cdot (x_k - x_i) & x_k^2 + y_k^2 - (x_i^2 + y_i^2) \end{bmatrix} \tag{8.11}$$

$$x_0 = \frac{\det(\mathbf{A})}{4((x_j - x_i)(y_k - y_i) - (x_k - x_i)(y_j - y_i))}$$

$$y_0 = \frac{\det(\mathbf{B})}{4((x_j - x_i)(y_k - y_i) - (x_k - x_i)(y_j - y_i))} \tag{8.12}$$

and

$$r = \sqrt{(x_0 - x_d)^2 + (y_0 - y_d)^2} \tag{8.13}$$

being det(.) the determinant and $d \in \{i, j, k\}$. Figure 8.3 illustrates the parameters defined by Eqs. (8.10–8.13).

Therefore it is possible to represent the shape parameters (for the circle, $[x_0, y_0, r]$) as a transformation R of the edge vector indexes i, j and k.

$$[x_0, y_0, r] = R(i, j, k) \tag{8.14}$$

with R being the transformation calculated after the previous computations of x_0, y_0, and r.

By exploring each index as an individual parameter, it is possible to sweep the continuous space looking for the shape parameters using the AIO through the CSA. This approach reduces the search space by eliminating unfeasible solutions.

Fig. 8.3 Circle candidate (individual) built from the combination of points p_i, p_j and p_k

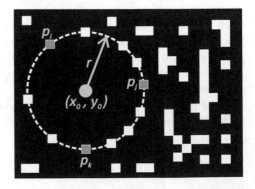

8.3.1.1 Objective Function or Matching Function

A circumference may be calculated as a virtual shape in order to measure the matching factor between C and the presented circle in the image (antigenic). It must be also validated, i.e. if it really exists in the edge image. The test for such points is $S = \{s_1, s_2, \ldots, s_{N_s}\}$, with N_s representing the number of test points over which the existence of an edge point will be verified.

The test S is generated by the midpoint circle algorithm (MCA) [44] which determines the required points for drawing a circle considering the radius r and the center point (x_0, y_0). The MCA employs the circle equation $x^2 + y^2 = r^2$ with only the first octant. It draws a curve starting at point $(r, 0)$ and proceeds upwards-left by using integer additions and subtractions. See full details in [45].

The MCA aims to calculate the points N_s which are required to represent the circle considering coordinates $S = \{s_1, s_2, \ldots, s_{N_s}\}$. Although the algorithm is considered the quickest providing a sub-pixel precision, it is important to assure that points lying outside the image plane must not be considered as they must be included in N_s, thus protecting the MCA operation.

The matching function or objective function $J(C)$ represents the matching (or error) resulting from pixels S for the circle candidate and the pixels that actually exist in the edge image, yielding:

$$J(C) = 1 - \frac{\sum_{i=1}^{N_s} E(x_i, y_i)}{N_s} \tag{8.15}$$

with $E(x_i, y_i)$ accumulating the number of expected edge points (the points in S) that are actually present in the edge image. N_s is the number of pixels within the perimeter of the circle that correspond to C, currently under testing.

Therefore the algorithm aims to minimize $J(C)$, given that a smaller value implies a better response (matching) of the "circularity" operator. The optimization process can thus be stopped after the maximum number of epochs is reached and the individuals are clearly defined satisfying the threshold. The stopping criterion depends on the a priori knowledge about the application context.

8.3.1.2 Implementation of CSA

In this work, an antibody will be represented (in binary form) by a bit chain of the form:

$$c = <c_1, c_2, \ldots, c_L> \tag{8.16}$$

where c representing a point in an L-dimensional (with L bits) space,

$$c \in S^L \tag{8.17}$$

The CSA implementation can be stated as follows:

1. An original pool of N antibodies is generated, considering the size of 22 bits.
2. The n best Ab's are selected based on the matching function. They will represent the memory set.
3. Best Ab's are cloned.
4. Perform hyper-mutation of the cloned Ab's following the affinity between antibodies and antigens while generating one improved antibody pool.
5. From the hyper-mutated pool, the Ab's with the highest affinity are to be re-selected.
6. As for the original pool, the Ab's with the lowest affinity are replaced improving the overall cells set.

Once the above steps are completed, the process is started again, until one Ab shows the best matching i.e. finding the minimum value of $J(C)$. In this work, the algorithm considers three index points embedded into a single Ab to represent one circle. Each single index has the variable P_i (with $i = 1, 2, 3$) representing the Hamming shape-space by means of a 22-bits word over the following range:

$$P_i : [1, N_p] \tag{8.18}$$

considering N_p the total number of edge pixels contained in the image. Hence, the first step is to generate the initial antibody pool by means of:

$$AB = 2 .* \text{rand}(N, S_p) - 1; \tag{8.19}$$

where S_p represents the bit size as it is assigned to each of N initial Abs –twenty two for this work. In order to perform the mapping from binary string to base 10, it yields

$$(\langle c_L, \ldots, c_2, c_1 \rangle)_2 = \left(\sum_{i=0}^{21} c_i \cdot 2^i \right)_{10} = r' \tag{8.20}$$

Finding the corresponding real value for r:

$$r = r' \cdot \frac{r_{max}}{2^{22} - 1} \tag{8.21}$$

by using r_{max} to represent N_p.

8.4 Experimental Results

8.4.1 Parametric Setup

Table 8.1 presents the parameters of CSA used in this work. Once they have been determined experimentally, they are kept for all the test images through all experiments.

All the experiments are performed on a Pentium IV 2.5 GHz computer under C language programming. All the images are preprocessed by the standard Canny edge-detector using the image-processing toolbox for MATLAB R2008a.

For comparison purposes, the CSA algorithm is tested against the IRHT and the GA circle detectors to each image individually.

For the GA algorithm described in Ayala-Ramirez et al. [22], the population size is 70, the crossover probability is 0.55, the mutation probability is 0.10 and number of elite individuals is 2. The roulette wheel selection and the 1-point crossover are applied. The parameter setup and the fitness function follow the configuration suggested in [22]. For the IRHT algorithm proposed in [19], the parameter values are defined as suggested in [19]. In IRHT, the most important parameters are grouped into the vector Δ_c which defines the desired set of enlargements of the circle/ellipse parameters to build a new region of interest. In this comparison, Δ_c is considered as $\Delta_c = [0.5 \cdot \sigma_x \quad 0.5 \cdot \sigma_x \quad 0.5 \cdot \sigma_a \quad 0.5 \cdot \sigma_b \quad 0]$. Such configuration is chose according to [19] as such values make the algorithm insensitive to noise images.

8.4.2 Error Score and Success Rate

Real-life images rarely contain perfectly-shaped circles. Therefore, in order to test the accuracy of the CSA approach, the results are compared to a ground-truth circle (see [23]) which is manually detected from the original edge-map. The parameters $(x_{true}, y_{true}, r_{true})$ of the ground-truth circle are computed using the Eqs. (8.10–8.13), over the three circumference points from the manually detected circle. If the center and the radius of such circle are found by the algorithm, defining (x_D, y_D) and r_D, then the error score defined as follows:

Table 8.1 Parameter setup for the CSA detector

h	n	N_c	ρ	P_r	L	T_e	*ITER*
120	100	80	10	20	22	0.01	400

$$Es = \eta \cdot (|x_{true} - x_D| + |y_{true} - y_D|) + \mu \cdot |r_{true} - r_D| \qquad (8.22)$$

The first term represents the shift of the center of the detected circle as it is compared to the ground-truth circle. The second term accounts for the difference between their radii. η and μ are two weights associated to each term in expression 8.22. They may be chosen according to the required accuracy as $\eta = 0.05$ and $\mu = 0.1$. This particular choice of parameters ensures that the radii difference is strongly weighted than difference of center positions between the manually detected and the machine-detected circle. It is assumed that if the Es is found to be less than 1, then the algorithm gets a success. Otherwise, it is considered to have failed in detecting the edge-circle. Notice that for $\eta = 0.05$ and $\mu = 0.1$ yields Es <1 which means that the maximum tolerated difference of radius length is 10 pixels while the maximum mismatch in the location of the center can be up to 20 pixels. From this viewpoint, the success rate (SR) is defined as percentage of reaching success after a certain number of trials.

8.4.2.1 Presentation of Results

Figure 8.4 provides three synthetic images and their counterparts after processing with CSA. Figure 8.5 presents the same experimental results on natural images. In

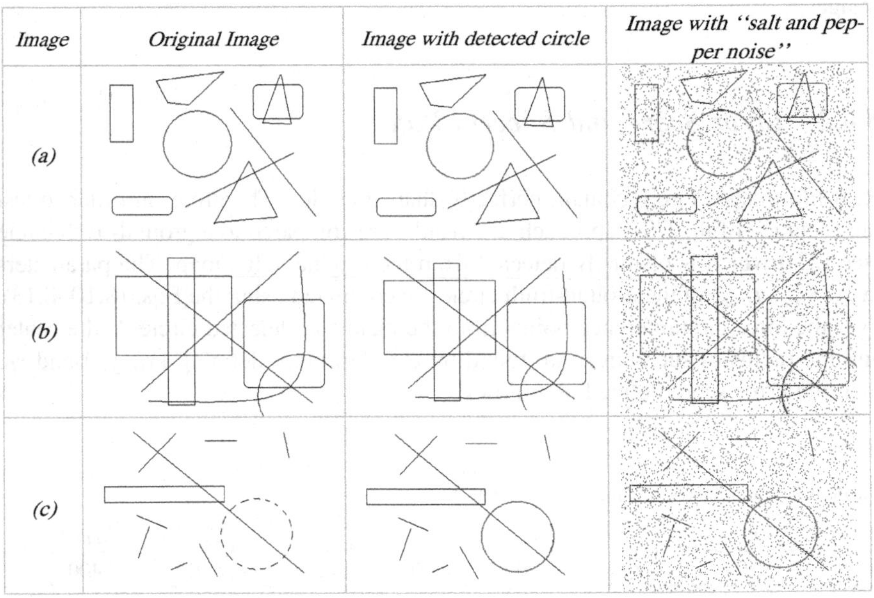

Image	Original Image	Image with detected circle	Image with "salt and pepper noise"
(a)			
(b)			
(c)			

Fig. 8.4 Synthetic images and their detected circles

order to test the robustness of the algorithm, salt and pepper noise have been added to the synthetic images before applying the algorithm. Likewise, the natural image shown by Fig. 8.5b is also corrupted with salt and pepper noise. It also illustrates the performance of the algorithm considering noisy and corrupted pixels. As real-life images rarely contain perfectly-shaped circles, the presented algorithm must approximate the circle that better fits into imperfect shapes within a noisy image. Such circle would therefore correspond to the better match in the objective function $J(C)$.

Considering the benchmark images and their corresponding edge maps shown by Figs. 8.4, 8.5 provides a visual performance illustration of the GA algorithm [22], the IRHT algorithm [19] and the presented approach, over three challenging problem instances, i.e. occluded circle, uneven circumference and synthetic noisy image).

The results are averaged over 35 independent runs of each algorithm. It is interesting to observe that the deviation between the detected circle and actual circle is the smallest under the CSA detector. Table 8.2 shows the averaged execution time, the success rate (in %), and averaged error score-following Eq. (8.22), for the three competitor algorithms over six test images shown by Figs. 8.4 and 8.5.

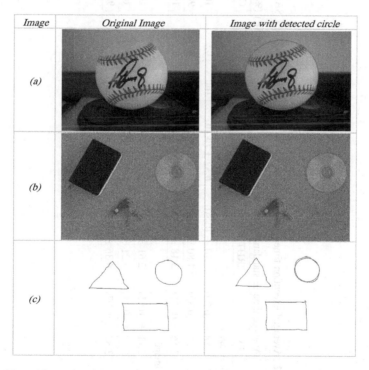

Image	Original Image	Image with detected circle
(a)		
(b)		
(c)		

Fig. 8.5 Natural images and their detected circles

Table 8.2 Averaged execution time and success rate of the GA, the IRHT and the CSA method, over the six test images shown by Figs. 8.4 and 8.5

Image	Average time ± standard deviation (s)			Success rate (SR) (%)			Es ± standard deviation		
	GA	IRHT	CSA	GA	IRHT	CSA	GA	IRHT	CSA
Synthetic images									
(a)	1.18 ± (0.20)	2.10 ± (0.80)	**0.52 ± (0.10)**	**100**	92	**100**	0.77 ± (0.081)	0.62 ± (0.070)	**0.40 ± (0.051)**
(b)	1.24 ± (0.39)	1.80 ± (0.65)	**0.46 ± (0.24)**	95	81	**98**	0.62 ± (0.050)	0.45 ± (0.023)	**0.37 ± (0.085)**
(c)	2.16 ± (0.11)	3.18 ± (0.36)	**0.60 ± (0.19)**	91	82	**100**	0.60 ± (0.041)	0.57 ± (0.041)	**0.31 ± (0.024)**
Natural images									
(a)	2.11 ± (0.51)	2.61 ± (0.52)	**1.12 ± (0.37)**	90	92	**100**	0.77 ± (0.031)	0.82 ± (0.043)	**0.43 ± (0.055)**
(b)	2.91 ± (0.34)	3.21 ± (0.14)	**1.61 ± (0.17)**	92	90	**100**	0.97 ± (0.055)	1.02 ± (0.136)	**0.51 ± (0.041)**
(c)	3.82 ± (0.97)	4.36 ± (0.17)	**1.95 ± (0.41)**	88	81	**98**	1.21 ± (0.102)	1.42 ± (0.155)	**0.59 ± (0.073)**

Table 8.3 Averaged execution time and success rate of the GA, the IRHT and the CSA method over three noisy images shown by Fig. 8.4

Image	Average time ± standard deviation (s)			Success rate (SR) (%)			Es ± standard deviation		
	GA	IRHT	CSA	GA	IRHT	CSA	GA	IRHT	CSA
Synthetic noisy images									
(a)	2.11 ± (0.31)	3.04 ± (0.29)	**0.57 ± (0.13)**	**100**	92	**100**	0.87 ± (0.071)	0.71 ± (0.051)	**0.54 ± (0.071)**
(b)	2.50 ± (0.39)	2.80 ± (0.17)	**0.51 ± (0.11)**	91	80	**97**	0.67 ± (0.081)	0.61 ± (0.048)	**0.31 ± (0.015)**
(c)	3.02 ± (0.63)	4.11 ± (0.71)	**0.64 ± (0.33)**	93	78	**100**	0.71 ± (0.036)	0.77 ± (0.044)	**0.42 ± (0.011)**

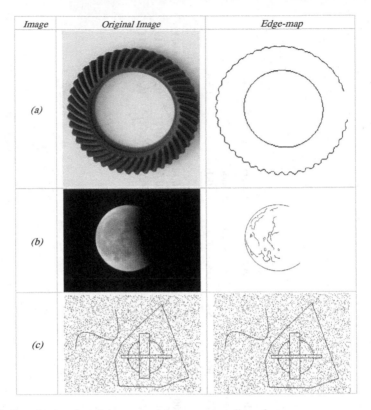

Image	Original Image	Edge-map
(a)		
(b)		
(c)		

Fig. 8.6 Complex benchmark images and their corresponding edge maps

Table 8.3 contents the results after processing noisy images shown by Fig. 8.4. The best results are marked in bold for both Tables.

The results are averaged over 35 independent runs of each algorithm. It is interesting to observe that the deviation between the detected circle and actual circle is the smallest under the CSA detector. Table 8.2 shows the averaged execution time, the success rate (in %), and averaged error score-following Eq. (8.22), for the three competitor algorithms over six test images shown by Figs. 8.4 and 8.5. Table 8.3 contents the results after processing noisy images shown by Fig. 8.4. The best results are marked in bold for both Tables.

A close inspection of Tables 8.2 and 8.3 reveals that the CSA method is able to achieve the highest success rate and minimum error tracking least computational time in majority of the cases.

Figure 8.6 shows some complex benchmark images, that were used to test the performance of the algorithms, the detection results for the three algorithms under consideration are shown below in Fig. 8.7.

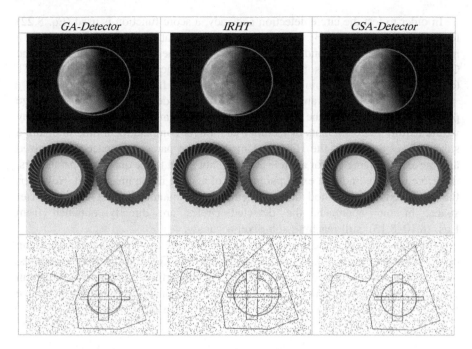

Fig. 8.7 Performance of the CSA method, the GA and the IRHT over complex images

8.5   Conclusions

This work has presented an algorithm for the automatic detection of circular shapes from complicated and noisy images with no consideration of the conventional Hough transform principles The presented method is based on a newly developed artificial immune optimization (AIO) technique, known as the clonal selection algorithm (CSA). To the best of our knowledge, the CSA has not been yet applied to any such circle detection task until date. The algorithm uses the encoding of three non-collinear edge points as circle candidates within the edge image of the scene. An objective function evaluates if a given circle candidate is actually present in the edge image (Ag). Guided by the values of the objective function, the set of encoded candidate circles are evolved using the CSA so that they can fit into the actual circles in the edge map of the image. As it can be observed from the results shown by Figs. 8.4, 8.5 and 8.7, our approach detects the circle in complex images with little visual distortion despite the presence of noisy background pixels.

An important feature is to consider the circle detection problem as an optimization approach. Such view enables the algorithm to detect arcs or occluded circles still matching imperfect circles. The CSA is capable of finding circle parameters according to $J(C)$ instead of making a review of all circle candidates towards detecting occluded or imperfect circles as it is commonly done by other methods.

In order to test the circle detection accuracy, a score function is used (Eq. 8.22) following the work in [23]. It can objectively evaluate the mismatch between a manually detected circle and a machine-detected shape. We demonstrated that the CSA method outperforms both the GA (as described in [22]) and the IRHT (as described in [19]) within a statistically significant framework.

Although the Hough Transform methods for circle detection also use three edge points to cast a vote for the potential circular shape in the parameter space, they would require huge amounts of memory and longer computational time to obtain a sub-pixel resolution. In the HT-based methods, the parameter space is quantized and the exact parameters for a circle are often not equal to the quantized parameters, therefore it rarely finds the exact parameters of a circle in the image [46]. However, the presented CSA method does not employ the quantization of the parameter space. In our approach, the detected circles are directly obtained from Eqs. (8.10–8.13), still reaching sub-pixel accuracy.

Although Fig. 8.6 indicates that the CSA method can yield better results on complicated and noisy images in comparison to the GA and the IRHT methods, notice that the aim of this chapter is to show that the Artificial Immune Systems can effectively serve as an attractive alternative to evolutionary algorithms which have been employed before to successfully extract circular shapes in images.

References

1. Brabazon A, O'Neill M (2006) Biologically inspired algorithms for financial modelling. Springer, Berlin
2. Chih-Chih L (2006) A novel image segmentation approach based on particle swarm optimization. IEICE Trans Fundam 89(1):324–327
3. Le Hégarat-Mascle S, Hégarat-Mascle L, Kallel A, Descombes X (2007) Ant colony optimization for image regularization based on a nonstationary markov modeling. IEEE Trans Image Process 16(3):865–878
4. Hammouche K, Diaf M, Siarry P (2008) Amultilevel automatic thresholding method based on a genetic algorithm for a fast image segmentation. Comput Vis Image Underst 109:163–175
5. Baştürk A, Günay E (2009) Efficient edge detection in digital images using a cellular neural network optimized by differential evolution algorithm. Expert Syst Appl 36(8):2645–2650
6. da Fontoura Costa L, Marcondes Cesar R Jr (2001) Shape Análisis and classification. CRC Press, Boca Raton
7. Peura M, Iivarinen J (1997) Efficiency of simple shape descriptors. In: Arcelli C, Cordella LP, di Baja GS (eds) Advances in visual form analysis. World Scientific, Singapore, pp 443–451
8. Yuen H, Princen J, Illingworth J, Kittler J (1990) Comparative study of Hough transform methods for circle finding. Image Vis Comput 8(1):71–77
9. Iivarinen J, Peura M, Sarela J, Visa A (1997) Comparison of combined shape descriptors for irregular objects. In: Proceedings of 8th British machine vision conference, Cochester, pp 430–439
10. Jones G, Princen J, Illingworth J, Kittler J (1990) Robust estimation of shape parameters. In: Proceedings of British machine vision conference, pp 43–48
11. Fischer M, Bolles R (1981) Random sample consensus: a paradigm to model fitting with applications to image analysis and automated cartography. CACM 24(6):381–395

12. Bongiovanni G, Crescenzi P (1995) Parallel simulated annealing for shape detection. Comput Vis Image Underst 61(1):60–69
13. Roth G, Levine MD (1994) Geometric primitive extraction using a genetic algorithm. IEEE Trans Pattern Anal Mach Intell 16(9):901–905
14. Muammar H, Nixon M (1989) Approaches to extending the Hough transform. In: Proceedings of international conference on acoustics, speech and signal processing ICASSP_89, vol 3, pp 1556–1559
15. Atherton TJ, Kerbyson DJ (1993) Using phase to represent radius in the coherent circle Hough transform. In: Proceedings of IEEE colloquium on the hough transform, IEE, London
16. Shaked D, Yaron O, Kiryati N (1996) Deriving stopping rules for the probabilistic Hough transform by sequential analysis. Comput Vis Image Underst 63:512–526
17. Xu L, Oja E, Kultanen P (1990) A new curve detection method: randomized Hough transform (RHT). Pattern Recognit Lett 11(5):331–338
18. Han JH, Koczy LT, Poston T (1993) Fuzzy Hough transform. In: Proceedings of 2nd international conference on fuzzy systems, vol 2, pp 803–808
19. Lu W, Tan JL (2008) Detection of incomplete ellipse in images with strong noise by iterative randomized Hough transform (IRHT). Pattern Recogn 41(4):1268–1279
20. Lutton E, Martinez P (1994) A genetic algorithm for the detection 2-D geometric primitives on images. In: Proceedings of the 12th international conference on pattern recognition, vol 1, pp 526–528
21. Yao J, Kharma N, Grogono P (2004) Fast robust GA-based ellipse detection. In: Proceedings of 17th international conference on pattern recognition ICPR-04, Cambridge, vol 2, pp 859–862
22. Ayala-Ramirez V, Garcia-Capulin CH, Perez-Garcia A, Sanchez-Yanez RE (2006) Circle detection on images using genetic algorithms. Pattern Recogn Lett 27:652–657
23. Dasgupta S, Das S, Biswas A, Abraham A (2009) Automatic circle detection on digital images whit an adaptive bacterial forganging algorithm. Soft Comput. doi:10.1007/s00500-009-0508-z
24. Rosin PL (1997) Further five point fit ellipse fitting. In: Proceedings of 8th British machine vision conference, Cochester, pp 290–299
25. Goldsby GA, Kindt TJ, Kuby J, Osborne BA (2003) Immunology, 5th edn. Freeman, New York
26. de Castro LN, Timmis J (2002) Artificial immune systems: a new computational intelligence approach. Springer, London
27. Dasgupta D (2006) Advances in artificial immune systems. IEEE Comput Intell Mag 1 (4):40–49
28. Wang X, Gao XZ, Ovaska SJ (2004) Artificial immune optimization methods and applications—a survey. In: Proceedings of the IEEE international conference on systems, man, and cybernetics, The Hague, pp 3415–3420
29. de Castro LN, von Zuben FJ (2002) Learning and optimization using the clonal selection principle. IEEE Trans Evol Comput 6(3):239–251
30. Ada GL, Nossal G (1987) The clonal selection theory. Sci Am 257:50–57
31. Coello Coello CA, Cortes NC (2005) Solving multiobjective optimization problems using an artificial immune system. Genet Program Evolvable Mach 6:163–190
32. Campelo F, Guimaraes FG, Igarashi H, Ramirez JA (2005) A clonal selection algorithm for optimization in electromagnetics. IEEE Trans Magn 41:1736–1739
33. Weisheng D, Guangming S, Li Z (2007) Immune memory clonal selection algorithms for designing stack filters. Neurocomputing 70:777–784
34. Gong M, Jiao L, Zhang L, Du H (2009) Immune secondary response and clonal selection inspired optimizers. Prog Nat Sci 19:237–253
35. de Castro LN, Member, IEEE, Von Zuben FJ, Member, IEEE (2002) Learning and optimization using the clonal selection principle. In: IEEE transactions on evolutionary computation, special issue on artificial immune systems, vol 6, no 3, pp 239–251

36. Cutello V, Narzisi G, Nicosia G, Pavone M (2005) Clonal selection algorithms: a comparative case study using effective mutation potentials. In: Jacob C et al (eds) ICARIS 2005, LNCS 3627, pp 13–28
37. Gong M, Jiao L, Zhang X (2008) A population-based artificial immune system for numerical optimization. Neurocomputing 72:149–161
38. Gao X, Wang X, Ovaska S (2009) Fusion of clonal selection algorithm and differential evolution method in training cascade-correlation neural network. doi:10.1016/j.neucom.2008.11.004
39. Poli R, Langdon WB (2002) Foundations of genetic programming. Springer, Berlin
40. Yoo J, Hajela P (1999) Immune network simulations in multicriterion design. Struct Optim 18 (2–3):85–94
41. Wang X, Gao XZ, Ovaska SJ (2005) A hybrid optimization algorithm in power filter design. In: Proceedings of the 31st annual conference of the IEEE industrial electronics society, Raleigh, November 2005, pp 1335–1340
42. Xu X, Zhang J (2007) An improved immune evolutionary algorithm for multimodal function optimization. In: Proceedings of the third international conference on natural computation, Haikou, August 2007, pp 641–646
43. Tang T, Qiu J (2006) An improved multimodal artificial immune algorithm and its convergence analysis. In: Proceedings of the sixth world congress on intelligent control and automation, Dalian, June 2006, pp 3335–3339
44. Bresenham JE (1987) A linear algorithm for incremental digital display of circular arcs. Commun ACM 20:100–106
45. Van Aken JR (1984) An efficient ellipse drawing algorithm. CG&A 4(9):24–35
46. Chen T-C, Chung K-L (2001) An eficient randomized algorithm for detecting circles. Comput Vis Image Underst 83:172–191

Chapter 9
States of Matter Algorithm Applied to Pattern Detection

Pattern Detection (PD) plays an important role in several image processing applications such as feature tracking, object recognition, stereo matching and remote sensing. PD involves two critical aspects: similarity measurement and search strategy. The simplest available PD method finds the best possible coincidence between the images through an exhaustive computation of the Normalized cross-correlation (NCC) values (similarity measurement) for all elements of the source image (search strategy). However, the use of such approach is strongly restricted, since the NCC evaluation is a computationally expensive operation. Recently, several PD algorithms, based on evolutionary approaches, have been proposed to reduce the number of NCC operations by calculating only a subset of search locations. In this chapter, is presented an algorithm based on the States of Matter with the purpose of reduce the number of search locations in the PD process. In the presented approach, individuals emulate molecules that experiment state transitions which represent different exploration–exploitation levels. In the algorithm, the computation of search locations is drastically reduced by incorporating a fitness calculation strategy which indicates when it is feasible to calculate or only estimate the NCC value for new search locations. Conducted simulations show that the presented method achieves the best balance over other PD algorithms, in terms of estimation accuracy and computational cost.

9.1 Introduction

Pattern Detection (PD), which measures the degree of similarity between two image sets that are superimposed on one another, is one of the most important and challenging subjects in digital photogrammetry, object recognition, stereo matching, feature tracking, remote sensing, and computer vision [1]. It relies on calculating at each position of the image under examination a correlation or distortion

© Springer International Publishing AG 2017   165
M.-A. Díaz-Cortés et al., *Engineering Applications of Soft Computing*,
Intelligent Systems Reference Library 129, DOI 10.1007/978-3-319-57813-2_9

function that measures the degree of similarity to the template image, and the best matching is obtained when the similarity value is maximized.

Generally, a process like pattern detection, involves two critical aspects: similarity measurement and search strategy [2]. It is used a matching criterion, typically the Normalized cross-correlation (NCC) which is computationally expensive and represents the most consuming operation in the PD process [3].

The full search algorithm [4–6] is the simplest PD algorithm that can deliver the optimal detection with respect to a maximal NCC coefficient as it checks all pixel-candidates one at a time. However, such exhaustive search and the NCC calculation at each checking point, yields an extremely computational expensive PD method that seriously constraints its use for several image processing applications.

Recently, several PD algorithms, based on evolutionary approaches, have been proposed to reduce the number of NCC operations by calculating only a subset of search locations. Such approaches have produced several robust detectors using different optimization methods such as Genetic algorithms (GA) [7], particle swarm optimization (PSO) [8, 9] and Imperialist competitive algorithm (ICA) [10]. Although these algorithms allow reducing the number of search locations, they do not explore the whole region effectively and often suffers premature convergence which conducts to sub-optimal detections. The reason of these problems is the operators used for modifying the particles. In such algorithms, during their evolution, the position of each agent in the next iteration is updated yielding an attraction towards the position of the best particle seen so-far [11, 12]. This behavior produces that the entire population, as the algorithm evolves, concentrates around the best particle, favoring the premature convergence and damaging the particle diversity.

Every evolutionary algorithm (EA) needs to address the issue of exploration-exploitation of the search space. Exploration is the process of visiting entirely new points of a search space whilst exploitation is the process of refining those points within the neighborhood of previously visited locations, in order to improve their solution quality. Pure exploration degrades the precision of the evolutionary process but increases its capacity to find new potential solutions. On the other hand, pure exploitation allows refining existent solutions but adversely driving the process to local optimal solutions. Therefore, the ability of an EA to find a global optimal solution depends on its capacity to find a good balance between the exploitation of found-so-far elements and the exploration of the search space [13]. So far, the exploration–exploitation dilemma has been an unsolved issue within the framework of EA.

In this chapter, a novel nature-inspired algorithm, called the States of Matter Search (SMS) is proposed for solving the PD problem. The SMS algorithm is based on the simulation of the states of matter phenomenon. In SMS, individuals emulate molecules which interact to each other by using evolutionary operations based on the physical principles of the thermal-energy motion mechanism. Such operations allow the increase of the population diversity and avoid the concentration of particles within a local minimum. The presented approach combines the use of the defined operators with a control strategy that modifies the parameter setting of each operation during the evolution process. The algorithm is devised by considering each state of matter at one different exploration–exploitation rate. Thus, the evolutionary process is divided into three stages which emulate the three states of matter: gas, liquid and solid. At each state, molecules (individuals) exhibit different behaviors. Beginning from the gas state (pure exploration), the algorithm modifies the intensities of exploration and exploitation until the solid state (pure exploitation) is reached. As a result, the approach can substantially improve the balance between exploration–exploitation, yet preserving the good search capabilities of an evolutionary approach.

However, one particular difficulty in applying any EA to real-world problems is about its demand for a large number of fitness evaluations before delivering a satisfying result. Fitness evaluations are not always straightforward in many applications as either an explicit fitness function does not exist or the fitness evaluation is computationally expensive. Furthermore, since random numbers are involved in the calculation of new individuals, they may encounter same positions (repetition) that have been visited by other individuals at previous iterations, particularly when individuals are confined to a finite area.

The problem of considering expensive fitness evaluations has already been faced in the field of evolutionary algorithms (EA) and is better known as fitness approximation [14]. In such approach, the idea is to estimate the fitness value of so many individuals as it is possible instead of evaluating the complete set. Such estimations are based on an approximate model of the fitness landscape. Thus, the individuals to be evaluated and those to be estimated are determined following some fixed criteria which depend on the specific properties of the approximate model [15]. The models involved at the estimation can be built during the actual EA run, since EA repeatedly samples the search space at different points [16]. There are many possible approximation models which have been used in combination with EA (e.g. polynomials [17], the kriging model [18], the feed-forward neural networks that includes multi-layer Perceptrons [19] and radial basis-function networks [20]).

In this chapter, a new algorithm based on SMS is presented to reduce the number of search locations in the PD process. The algorithm uses a simple fitness calculation approach which is based on the Nearest Neighbor Interpolation (NNI) algorithm in order to estimate the fitness value (NCC operation) for several candidate solutions (search locations). As a result, the approach can not only substantially reduce the number search positions (by using the SMS approach), but also to avoid the NCC evaluation for many of them (by incorporating the NNI strategy). The presented method achieves the best balance over other PD algorithms, in terms of both estimation accuracy and computational cost.

The overall chapter is organized as follows: Sect. 9.2 holds a description about the SMS algorithm. In Sect. 9.3, the fitness calculation strategy for solving the expensive optimization problem is presented. Section 9.4 provides backgrounds about the PD process while Sect. 9.5 exposes the final PD algorithm as a combination of SMS and the fitness calculation strategy. Section 9.6 demonstrates experimental results for the presented approach over standard test images and some conclusions are drawn in Sect. 9.7.

9.2 Gaussian Approximation

The matter can take different phases which are commonly known as states. Traditionally, three states of matter are known: solid, liquid, and gas. The differences among such states are based on forces which are exerted among particles composing a material [21].

In the gas phase, molecules present enough kinetic energy so that the effect of intermolecular forces is small (or zero for an ideal gas), while the typical distance between neighboring molecules is greater than the molecular size. A gas has no definite shape or volume, but occupies the entire container in which it is confined. Figure 9.1a shows the movements exerted by particles in a gas state. The movement experimented by the molecules represent the maximum permissible displacement ρ_1 among particles [22]. In a liquid state, intermolecular forces are more restrictive than those in the gas state. The molecules have enough energy to move relatively to each other still keeping a mobile structure. Therefore, the shape of a liquid is not definite but is determined by its container. Figure 9.1b presents a particle movement ρ_2 within a liquid state. Such movement is smaller than those considered by the gas state but larger than the solid state. In the solid state, particles (or molecules) are packed together closely with forces among particles being strong enough so that the particles cannot move freely but only vibrate. As a result, a solid has a stable, definite shape and a definite volume. Solids can only change their shape by force, as when they are broken or cut. Figure 9.1c shows a molecule configuration in a solid state. Under such conditions, particles are able to vibrate (being perturbed) considering a minimal ρ_3 distance [22].

In this chapter, a novel nature-inspired algorithm known as the States of Matter Search (SMS) is presented for solving global optimization problems. The SMS algorithm is based on the simulation of the states of matter phenomenon that

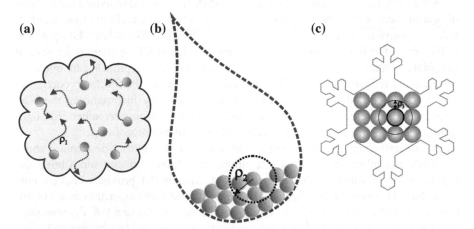

Fig. 9.1 Different states of matter: **a** gas, **b** liquid, and **c** solid

considers individuals as molecules which interact to each other by using evolutionary operations based on the physical principles of the thermal-energy motion mechanism. The algorithm is devised by considering each state of matter at one different exploration–exploitation ratio. Thus, the evolutionary process is divided into three stages which emulate the three states of matter: gas, liquid and solid. In each state, individuals exhibit different behaviors.

9.3 States of Matter Search (SMS)

9.3.1 Definition of Operators

In the approach, individuals are considered as molecules whose positions on a multidimensional space are modified as the algorithm evolves. The movement of such molecules is motivated by the analogy to the motion of thermal-energy.

The velocity and direction of each molecule's movement are determined by considering the collision, the attraction forces and the random phenomena experimented by the molecule set [23]. In our approach, such behaviors have been implemented by defining several operators such as the direction vector, the collision and the random positions operators, all of which emulate the behavior of actual physics laws.

The direction vector operator assigns a direction to each molecule in order to lead the particle movement as the evolution process takes place. On the other side, the collision operator mimics those collisions that are experimented by molecules as they interact to each other. A collision is considered when the distance between two molecules is shorter than a determined proximity distance. The collision operator is thus implemented by interchanging directions of the involved molecules. In order to simulate the random behavior of molecules, the proposed algorithm generates random positions following a probabilistic criterion that considers random locations within a feasible search space.

The next section presents all operators that are used in the algorithm. Although such operators are the same for all the states of matter, they are employed over a different configuration set depending on the particular state under consideration.

9.3.1.1 Direction Vector

The direction vector operator mimics the way in which molecules change their positions as the evolution process develops. For each n-dimensional molecule $\mathbf{p}_i$ from the population $\mathbf{P}$, it is assigned an n-dimensional direction vector $\mathbf{d}_i$ which stores the vector that controls the particle movement. Initially, all the direction vectors $\left(\mathbf{D} = \{\mathbf{d}_1, \mathbf{d}_2, \ldots, \mathbf{d}_{N_p}\}\right)$ are randomly chosen within the range of $[-1, 1]$.

As the system evolves, molecules experiment several attraction forces. In order to simulate such forces, the proposed algorithm implements the attraction phenomenon by moving each molecule towards the best so-far particle. Therefore, the new direction vector for each molecule is iteratively computed considering the following model:

$$\mathbf{d}_i^{k+1} = \mathbf{d}_i^k \cdot \left(1 - \frac{k}{gen}\right) \cdot 0.5 + \mathbf{a}_i, \tag{9.1}$$

where $\mathbf{a}_i$ represents the attraction unitary vector calculated as $\mathbf{a}_i = (\mathbf{p}^{best} - \mathbf{p}_i)/\|\mathbf{p}^{best} - \mathbf{p}_i\|$, being $\mathbf{p}^{best}$ the best individual seen so-far, while $\mathbf{p}_i$ is the molecule i of population $\mathbf{P}$. k represents the iteration number whereas gen involves the total iteration number that constitutes the complete evolution process.

Under this operation, each particle is moved towards a new direction which combines the past direction, which was initially computed, with the attraction vector over the best individual seen so-far. It is important to point out that the relative importance of the past direction decreases as the evolving process advances. This particular type of interaction avoids the quick concentration of information among particles and encourages each particle to search around a local candidate region in its neighborhood, rather than interacting to a particle lying at distant region of the domain. The use of this scheme has two advantages: first, it prevents the particles from moving toward the global best position in early stages of algorithm and thus makes the algorithm less susceptible to premature convergence; second, it encourages particles to explore their own neighborhood thoroughly, just before they converge towards a global best position. Therefore, it provides the algorithm with local search ability enhancing the exploitative behavior.

In order to calculate the new molecule position, it is necessary to compute the velocity $\mathbf{v}_i$ of each molecule by using:

$$\mathbf{v}_i = \mathbf{d}_i \cdot v_{init} \tag{9.2}$$

being v_{init} the initial velocity magnitude which is calculated as follows:

$$v_{init} = \frac{\sum_{j=1}^{n}(b_j^{high} - b_j^{low})}{n} \cdot \beta \tag{9.3}$$

where b_j^{low} and b_j^{high} are the low j parameter bound and the upper j parameter bound respectively, whereas $\beta \in [0, 1]$.

Then, the new position for each molecule is updated by:

$$p_{i,j}^{k+1} = p_{i,j}^k + v_{i,j} \cdot \text{rand}(0, 1) \cdot \rho \cdot (b_j^{high} - b_j^{low}) \tag{9.4}$$

where $0.5 \leq \rho \leq 1$.

9.3.1.2 Collision

The collision operator mimics the collisions experimented by molecules while they interact to each other. Collisions are calculated if the distance between two molecules is shorter than a determined proximity value. Therefore, if $\|\mathbf{p}_i - \mathbf{p}_q\| < r$, a collision between molecules i and q is assumed; otherwise, there is no collision, considering $i, q \in \{1, \ldots, N_p\}$ such that $i \neq q$. If a collision occurs, the direction vector for each particle is modified by interchanging their respective direction vectors as follows:

$$\mathbf{d}_i = \mathbf{d}_q \text{ and } \mathbf{d}_q = \mathbf{d}_i \tag{9.5}$$

The collision radius is calculated by:

$$r = \frac{\sum_{j=1}^{n} \left(b_j^{high} - b_j^{low} \right)}{n} \cdot \alpha \tag{9.6}$$

where $\alpha \in [0, 1]$.

Under this operator, a spatial region enclosed within the radius r is assigned to each particle. In case the particle regions collide to each other, the collision operator acts upon particles by forcing them out of the region. The radio r and the collision operator provide the ability to control diversity throughout the search process. In other words, the rate of increase or decrease of diversity is predetermined for each stage. Unlike other diversity-guided algorithms, it is not necessary to inject diversity into the population when particles gather around a local optimum because the diversity will be preserved during the overall search process. The collision incorporation therefore enhances the exploratory behavior in the proposed approach.

9.3.1.3 Random Positions

In order to simulate the random behavior of molecules, the proposed algorithm generates random positions following a probabilistic criterion within a feasible search space.

For this operation, a uniform random number r_m is generated within the range [0, 1]. If r_m is smaller than a threshold H, a random molecule's position is generated; otherwise, the element remains with no change. Therefore such operation can be modeled as follows:

$$p_{i,j}^{k+1} = \begin{cases} b_j^{low} + rand(0, 1) \cdot \left(b_j^{high} - b_j^{low} \right) & \text{with probability } H \\ p_{i,j}^{k+1} & \text{with probability } (1 - H) \end{cases} \tag{9.7}$$

where $i \in \{1, \ldots, N_p\}$ and $j \in \{1, \ldots, n\}$.

9.3.1.4   Best Element Updating

Despite this updating operator does not belong to State of Matter metaphor, it is used to simply store the best so-far solution. In order to update the best molecule $\mathbf{p}^{best}$ seen so-far, the best found individual from the current k population $\mathbf{p}^{best,k}$ is compared to the best individual $\mathbf{p}^{best,k-1}$ of the last generation. If $\mathbf{p}^{best,k}$ is better than $\mathbf{p}^{best,k-1}$ according to its fitness value, $\mathbf{p}^{best}$ is updated with $\mathbf{p}^{best,k}$, otherwise $\mathbf{p}^{best}$ remains with no change. Therefore, $\mathbf{p}^{best}$ stores the best historical individual found so-far.

9.3.2   SMS Algorithm

The overall SMS algorithm is composed of three stages corresponding to the three States of Matter: the gas, the liquid and the solid state. Each stage has its own behavior. In the first stage (gas state), exploration is intensified whereas in the second one (liquid state) a mild transition between exploration and exploitation is executed. Finally, in the third phase (solid state), solutions are refined by emphasizing the exploitation process.

9.3.2.1   General Procedure

At each stage, the same operations are implemented. However, depending on which state is referred, they are employed considering a different parameter configuration. The general procedure in each state is shown as pseudo-code in Algorithm 9.1. Such procedure is composed by five steps and maps the current population $\mathbf{P}^k$ to a new population $\mathbf{P}^{k+1}$. The algorithm receives as input the current population $\mathbf{P}^k$ and the configuration parameters ρ, β, α, and H, whereas it yields the new population $\mathbf{P}^{k+1}$.

9.3.2.2   The Complete Algorithm

The complete algorithm is divided into four different parts. The first corresponds to the initialization stage, whereas the last three represent the States of Matter. All the optimization process, which consists of a *gen* number of iterations, is organized into three different asymmetric phases, employing 50% of all iterations for the gas state (exploration), 40% for the liquid state (exploration-exploitation) and 10% for the solid state (exploitation). The overall process is graphically described by Fig. 9.2. At each state, the same general procedure (see Algorithm 9.1) is iteratively used considering the particular configuration predefined for each State of Matter. Figure 9.3 shows the data flow for the complete SMS algorithm.

Algorithm 9.1. General procedure executed by all the states of matter

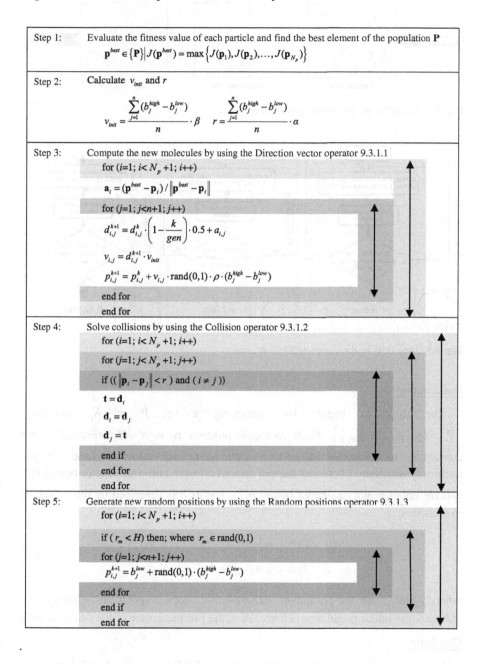

Step 1:	Evaluate the fitness value of each particle and find the best element of the population **P** $$\mathbf{p}^{best} \in \{\mathbf{P}\} \mid J(\mathbf{p}^{best}) = \max\{J(\mathbf{p}_1), J(\mathbf{p}_2), \ldots, J(\mathbf{p}_{N_p})\}$$
Step 2:	Calculate v_{init} and r $$v_{init} = \frac{\sum_{j=1}^{n}(b_j^{high} - b_j^{low})}{n} \cdot \beta \qquad r = \frac{\sum_{j=1}^{n}(b_j^{high} - b_j^{low})}{n} \cdot \alpha$$
Step 3:	Compute the new molecules by using the Direction vector operator 9.3.1.1 for $(i=1; i< N_p +1; i++)$ $$\mathbf{a}_i = (\mathbf{p}^{best} - \mathbf{p}_i)/\|\mathbf{p}^{best} - \mathbf{p}_i\|$$ for $(j=1; j<n+1; j++)$ $$d_{i,j}^{k+1} = d_{i,j}^{k} \cdot \left(1 - \frac{k}{gen}\right) \cdot 0.5 + a_{i,j}$$ $$v_{i,j} = d_{i,j}^{k+1} \cdot v_{init}$$ $$p_{i,j}^{k+1} = p_{i,j}^{k} + v_{i,j} \cdot \text{rand}(0,1) \cdot \rho \cdot (b_j^{high} - b_j^{low})$$ end for end for
Step 4:	Solve collisions by using the Collision operator 9.3.1.2 for $(i=1; i< N_p +1; i++)$ for $(j=1; j< N_p +1; j++)$ if $((\|\mathbf{p}_i - \mathbf{p}_j\| < r)$ and $(i \neq j))$ $\mathbf{t} = \mathbf{d}_i$ $\mathbf{d}_i = \mathbf{d}_j$ $\mathbf{d}_j = \mathbf{t}$ end if end for end for
Step 5:	Generate new random positions by using the Random positions operator 9.3.1.3 for $(i=1; i< N_p +1; i++)$ if $(r_m < H)$ then; where $r_m \in \text{rand}(0,1)$ for $(j=1; j<n+1; j++)$ $$p_{i,j}^{k+1} = b_j^{low} + \text{rand}(0,1) \cdot (b_j^{high} - b_j^{low})$$ end for end if end for

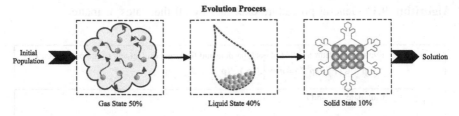

Fig. 9.2 Evolution process in the proposed approach

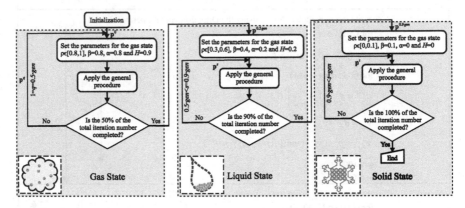

Fig. 9.3 Data flow in the complete SMS algorithm

Initialization

The algorithm begins by initializing a set $\mathbf{P}$ of N_p molecules $\left(\mathbf{P} = \{\mathbf{p}_1, \mathbf{p}_2, \ldots, \mathbf{p}_{N_p}\}\right)$. Each molecule position $\mathbf{p}_i$ is a n-dimensional vector containing the parameter values to be optimized. Such values are randomly and uniformly distributed between the pre-specified lower initial parameter bound b_j^{low} and the upper initial parameter bound b_j^{high}, just as it is described by the following expressions:

$$p_{i,j}^0 = b_j^{low} + \text{rand}(0,1) \cdot (b_j^{high} - b_j^{low})$$
$$j = 1, 2, \ldots, n; \quad i = 1, 2, \ldots, N_p, \tag{9.8}$$

where j and i, are the parameter and molecule index respectively whereas zero indicates the initial population. Hence, p_i^j is the j-th parameter of the i-th molecule.

Gas State

In the gas state, molecules experiment severe displacements and collisions. Such state is characterized by random movements produced by non-modeled molecule phenomena [23]. Therefore, the ρ value from the direction vector operator is set to a value near to one so that the molecules can travel longer distances. Similarly, the H value representing the random positions operator is also configured to a value around one, in order to allow the random generation for other molecule positions. The gas state is the first phase and lasts for the 50% of all iterations which compose the complete optimization process. The computational procedure for the gas state can be summarized as follows:

Step 1: Set the parameters $\rho \in [0.8, 1]$, $\beta = 0.8$, $\alpha = 0.8$, and $H = 0.9$ being consistent with the gas state.
Step 2: Apply the general procedure which is illustrated in Algorithm 9.1.
Step 3: If the 50% of the total iteration number is completed $(1 \leq k \leq 0.5 \cdot gen)$, then the process continues to the liquid state procedure; otherwise go back to step 2.

Liquid State

Although molecules currently at the liquid state exhibit restricted motion in comparison to the gas state, they still show a higher flexibility with respect to the solid state. Furthermore, the generation of random positions which are produced by non-modeled molecule phenomena is scarce [24]. For this reason, the ρ value from the direction vector operator is bounded to a value between 0.3 and 0.6. Similarly, the random position operator H is configured to a value near to cero in order to allow the random generation of fewer molecule positions. In the liquid state, collisions are also less common than in gas state, so the collision radius, that is controlled by α, is set to a smaller value in comparison to the gas state. The liquid state is the second phase and lasts the 40% of all iterations which compose the complete optimization process. The computational procedure for the liquid state can be summarized as follows:

Step 4: Set the parameters $\rho \in [0.3, 0.6]$, $\beta = 0.4$, $\alpha = 0.2$ and $H = 0.2$ being consistent with the liquid state.
Step 5: Apply the general procedure that is defined in Algorithm 9.1.
Step 6: If the 90% (50% from the gas state and 40% from the liquid state) of the total iteration number is completed $(0.5 \cdot gen < k \leq 0.9 \cdot gen)$, then the process continues to the solid state procedure; otherwise go back to step 5.

Solid State

In the solid state, forces among particles are stronger so that particles cannot move freely but only vibrate. As a result, effects such as collision and generation of random positions are not considered [25]. Therefore, the ρ value of the direction vector operator is set to a value near to zero indicating that the molecules can only vibrate around their original positions. The solid state is the third phase and lasts for the 10% of all iterations which compose the complete optimization process. The computational procedure for the solid state can be summarized as follows:

Step 7: Set the parameters $\rho \in [0.0, 0.1]$ and $\beta = 0.1, \alpha = 0$ and $H = 0$ being consistent with the solid state.

Step 8: Apply the general procedure that is defined in Algorithm 9.1.

Step 9: If the 100% of the total iteration number is completed ($0.9 \cdot gen < k \leq gen$), the process is finished; otherwise go back to step 8.

It is important to clarify that the use of this particular configuration ($\alpha = 0$ and $H = 0$) disables the collision and generation of random positions operators which have been illustrated in the general procedure.

9.4   Fitness Approximation Method

Evolutionary methods based on fitness approximation aim to find the global optimum of a given function considering only a very few number of function evaluations. In order to apply such approach, it is necessary that the objective function portrait the following conditions: [18]: (1) it must be very costly to evaluate and (2) must have few dimensions (up to five). Recently, several fitness estimators have been reported in the literature [14–17], where the function evaluation number is considerably reduced (to hundreds, dozens, or even less). However, most of these methods produce complex algorithms whose performance is conditioned to the quality of the training phase and the learning algorithm in the construction of the approximation model.

In this chapter, we explore the use of a local approximation scheme, based on the nearest-neighbor-interpolation (NNI), in order to reduce the function evaluation number. The model estimates the fitness values based on previously evaluated neighboring individuals, stored during the evolution process. In each generation, some individuals of the population are evaluated with the accurate (real) objective function, while the remaining individuals' fitnesses are estimated. The individuals to be evaluated accurately are determined based on their proximity to the best fitness value or uncertainty.

9.4.1   Updating Individual Database

In our fitness calculation approach, during de evolution process, every evaluation or estimation of an individual produces a data point (individual position and fitness value) that is potentially taken into account for building the approximation model. Therefore, we keep all seen so far evaluations in a history array $\mathbf{T}$, and then just select the closest neighbor to estimate the fitness value of a new individual. Thus, all data are preserved and potentially available for use, while the construction of the model is still fast since only the most relevant data points are actually used to construct the model.

9.4.2 Fitness Calculation Strategy

In the presented fitness calculation scheme, most of the fitness values are estimated to reduce the calculation time in each generation. In the model, it is evaluated (using the real fitness function) those individuals that are near the individual with the best fitness value contained in **T** (rule 1). Such individuals are important, since they will have a stronger influence on the evolution process than other individuals. Moreover, it is also evaluated those individuals in regions of the search space with few previous evaluations (rule 2). The fitness values of these individuals are uncertain; since there is no close reference (close points contained in **T**) in order to calculate their estimates.

The rest of the individuals are estimated using NNI (rule 3). Thus, the fitness value of an individual is estimated assigning it the same fitness value that the nearest individual stored in **T**. For the sake of clarity, it is considered that the fitness value of i is evaluated by the true fitness function using the representation J (i) whereas $\tilde{J}(i)$ indicates that the fitness value of the individual i has been estimated using an alternative model.

Therefore, the estimation model follows 3 different rules in order to evaluate or estimate the fitness values:

1. If the new individual (search position) P is located closer than a distance d with respect to the nearest individual location L_q whose fitness value F_{L_q} corresponds to the best fitness value stored in **T**, then the fitness value of P is evaluated using the true fitness function ($J(P)$). Figure 9.4a draws the rule procedure.
2. If the new individual P is located longer than a distance d with respect to the nearest individual location L_q whose fitness value F_{L_q} has been already stored in **T**, then its fitness value is evaluated using the true fitness function [$J(P)$]. Figure 9.4b outlines the rule procedure.
3. If the new individual P is located closer than a distance d with respect to the nearest individual location L_q whose fitness value F_{L_q} has been already stored in **T**, then its fitness value is estimated $(\tilde{J}(P))$ assigning it the same fitness that $L_q (F_P = F_{L_q})$. Figure 9.4c sketches the rule procedure.

The d value controls the trade off between the evaluation and estimation of search locations. Typical values of d range from 5 to 10; however, in this chapter, the value of 7 has been selected. Thus, the proposed approach favors the exploitation and exploration in the search process. For the exploration, the estimator evaluates the true fitness function of new search locations that have been located far

Fig. 9.4 The fitness
calculation strategy.
a According to the rule 1, the
individual (search position)
P is evaluated ($J(P)$), since it
is located closer than a
distance d with respect to the
nearest individual location L_1
whose fitness value F_{L_1}
corresponds to the best fitness
value (maximum so-far),
b according to the rule 2, the
search point P is evaluated (J
(P)), as there is no close
reference in its neighborhood
and **c** according to the rule 3,
the fitness value of p is
estimated ($\tilde{J}(P)$) by means of
the NNI-estimator, assigning
$F_P = F_{L_2}$

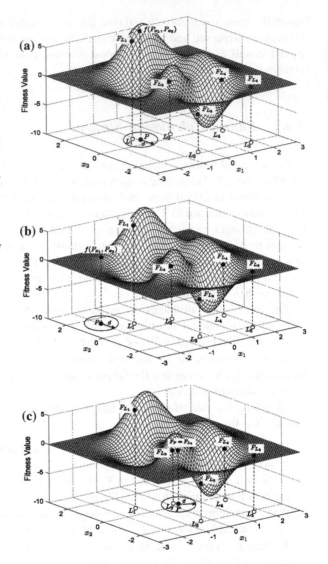

from the positions already calculated. Meanwhile, it also estimates those that are
closer. For the exploitation, the proposed method evaluates the effective fitness
function of those new searching locations that are placed near to the position with
the minimum fitness value seen so far, aiming to improve its minimum.

The three rules show that the fitness calculation strategy is simple and
straightforward. Figure 9.4 illustrates the procedure of fitness computation for a
new solution (point P) considering the three different rules. In the problem, the
objective function J is maximized with respect to two parameters (x_1, x_2). In all
figures (Fig. 9.4a–c) the individual database array **T** contains five different elements

$(L_1, L_2, L_3, L_4, L_5)$ with their corresponding fitness values $(F_{L_1}, F_{L_2}, F_{L_3}, F_{L_4}, F_{L_5})$. Figure 9.4a, b show the fitness evaluation $(J(P_{x_1}, P_{x_2}))$ of the new solution P following the rule 1 and 2 respectively, whereas Fig. 9.4c present the fitness estimation of P $(\tilde{J}(P))$ using the NNI approach considered by rule 3.

Algorithm 9.2 Enhanced general procedure executed by all the states of matter. The procedure incorporates the fitness calculation strategy in order to reduce the number of function evaluations

Step 1:	Evaluate or estimate the fitness value of each particle and find the best element of the population **P**
	for $(i=1; i< N_p +1; i{+}{+})$
	If ($\mathbf{p}_i$ fulfils rule 1 or rule 2) then $J(\mathbf{p}_i)$
	If ($\mathbf{p}_i$ fulfils rule 3) then $\tilde{J}(\mathbf{p}_i)$
	update **T**
	end for
	$\mathbf{p}^{best} \in \{\mathbf{T}\} \big| J(\mathbf{p}^{best}) = \max\left\{ J(\mathbf{p}_1), \tilde{J}(\mathbf{p}_2), \ldots, J(\mathbf{p}_{N_p}) \right\}$
Step 2:	Calculate v_{init} and r
	$$v_{init} = \frac{\sum_{j=1}^{n}(b_j^{high} - b_j^{low})}{n} \cdot \beta \qquad r = \frac{\sum_{j=1}^{n}(b_j^{high} - b_j^{low})}{n} \cdot \alpha$$
	The other steps are similar to those presented in Algorithm 1.

9.4.3 Proposed Optimization SMS Method

In this section, it has been proposed a fitness calculation approach in order to accelerate the SMS algorithm. Only the fitness calculation scheme shows difference between the proposed SMS and the enhanced one. In the modified SMS, only some individuals are actually evaluated (rules 1 and 2) in each generation. The fitness values of the rest are estimated using the NNI-approach (rule 3). The estimation is executed using the individual database (array **T**).

Figure 9.5 shows the difference between the original SMS and the modified one. In the figure, it is clear that two new blocks have been added, the fitness estimation and the updating individual database. Both elements, together with the actual evaluation block, represent the fitness calculation strategy presented in this sub-section. The incorporation of the fitness calculation strategy modifies only the step 1 of the general procedure shown in Algorithm 9.1. Such step is extended by incorporating the decision rules [whether the individual i is $J(i)$ or $\tilde{J}(i)$) and the

Fig. 9.5 Differences between
the original SMS and the
modified SMS.
a Conventional SMS and
b SMS algorithm included the
fitness calculation strategy

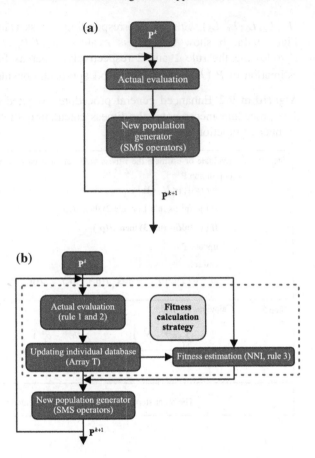

sub-system that updates the **T** array. Algorithm 9.2 illustrates the enhanced procedure. As a result, the SMS approach can substantially reduce the number of function evaluations preserving its good search capabilities.

9.5   Pattern Detection Process

Consider the problem of localizing a given reference image (template) R within a larger intensity image I, which we call the source image. The task is to find those positions where the contents of the reference image R and the corresponding sub-image of I are either the same or most similar. If it is denoted by $R_{u,v}(x, y) = R(x - u, y - v)$, the reference image R shifted by the distance (u, v) in the horizontal and vertical directions, respectively, then the matching problem

Fig. 9.6 Geometry of pattern detection. The reference image R is shifted across the search image I by an offset (u, v) using the origins of the two images as the reference points. The dimensions of the source image (MxN) and the reference image ($m \times n$) determine the maximal search region (S) for this comparison

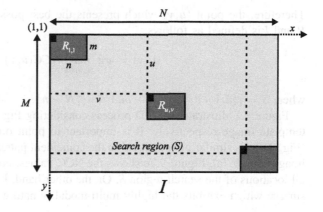

(illustrated in Fig. 9.6) can be summarized as: given are the source image I and the reference image R. Find the offset (u, v) inside of the search region S, such that the similarity between the shifted reference image $T_{u,v}$ and the corresponding sub-image of I is a maximum.

To successfully solve this task, several issues need to be addressed such as determining a minimum similarity value for accepting a match and developing a good search strategy for finding, in a fast way, the optimal displacement. Several Pattern Detection algorithms [7–10] have been proposed to reduce the number of search positions, using evolutionary approaches as search strategy. Among the similarity criteria, NCC is the most effective and robust method that allow to measure the resemblance between R and its coincident region within I, at each displacement (u, v). The NCC value between a given image I of size $M \times N$, and a template image R of size $m \times n$ at the displacement (u, v) is given by:

$$NCC(u, v) = \frac{\sum_{i=1}^{m} \sum_{j=1}^{n} [I(u+i, v+j) - \bar{I}(u, v)] \cdot [R(i, j) - \bar{R}]}{\left[\sum_{i=1}^{m} \sum_{j=1}^{n} I(u+i, v+j) - \bar{I}(u, v) \right]^{\frac{1}{2}} \cdot \left[\sum_{i=1}^{m} \sum_{j=1}^{n} R(i, j) - \bar{R} \right]^{\frac{1}{2}}}$$

$$(9.9)$$

where $\bar{I}(u, v)$ is the grey-scale average intensity of the source image in the region coincident with the template image R and $\bar{R}$ is the grey-scale average intensity of the template image. Such values are defined as:

$$\bar{I}(u, v) = \frac{1}{m \cdot n} \sum_{i=1}^{m} \sum_{j=1}^{n} I(u+i, v+j)$$

$$\bar{R} = \frac{1}{m \cdot n} \sum_{i=1}^{m} \sum_{j=1}^{m} R(i, j)$$

$$(9.10)$$

Therefore, the point (u, v) which presents the best possible resemblance between R and I is defined as follows:

$$(u, v) = \arg \max_{(\hat{u},\hat{v}) \in S} NCC(\hat{u}, \hat{v}) \tag{9.11}$$

where $S = \{(\hat{u}, \hat{v}) \mid 1 \leq \hat{u} \leq M - m, 1 \leq \hat{v} \leq N - n\}$.

Figure 9.7 illustrates the PD process considering Fig. 9.7a, b as the source and template image respectively. It is important to point out that the template image (Fig. 9.7b) is similar but not equal to the coincident pattern, contained in the source image (Fig. 9.7a). Figure 9.7c shows the NCC values (color-encoded) calculated in all locations of the search region S. On the other hand, Fig. 9.7d presents the NCC surface which exhibits the highly multi-modality nature of the PD problem.

9.6   PD Algorithm Based on SMS with the Estimation Strategy

The simplest available PD method finds the global maximum (the accurate detection point (u, v)), considering all locations within the search space S. Nevertheless, the approach has a high computational cost for its practical use. Several PD algorithms [7–10] have been proposed to accelerate the search process by calculating only a subset of search locations. Although these algorithms allow reducing the number of search locations, they do not explore the whole region effectively and often suffers premature convergence which conducts to sub-optimal detections. The cause of these problems is the operators used for modifying the particles. In such algorithms, during their evolution, the position of each agent in the next iteration is updated yielding an attraction towards the position of the best particle seen so-far [19, 20]. This behavior produces that the entire population, as the algorithm evolves, concentrates around the best coincidence seen so-far, favoring the premature convergence in a local minima of the multi-modal surface. Therefore, a better PD algorithm should spend less computational time on the search strategy and get the optimum match position.

In the SMS-based algorithm, individuals represent search positions (u, v) which move throughout the search space S. The NCC coefficient, used as a fitness value, evaluates the matching quality presented between the template image R and the source image I, for a determined search position (individual). The number of NCC evaluations is drastically reduced by considering a fitness calculation strategy which indicates when it is feasible to calculate or only estimate the NCC values for new search locations. Guided by the fitness values (NCC coefficients), the set of encoded candidate positions are evolved using the SMS operators until the best possible resemblance has been found.

In the algorithm, the search space S consists of a set of 2-D search positions $\hat{u}$ and $\hat{v}$ representing the x and y components of the detection locations, respectively. Each particle is thus defined as:

$$P_i = \{(\hat{u}_i, \hat{v}_i)| \, 1 \le \hat{u}_i \le M - m, 1 \le \hat{v}_i \le N - n\} \qquad (9.12)$$

9.6.1 The SMS-PD Algorithm

The goal of our PD-approach is to reduce the number of evaluations of the NCC values (actual fitness function) avoiding any performance loss and achieving the optimal solution. The SMS-PD method is listed below:

Step 1:   Set the SMS parameters.

Step 2:   Initialize the population of 5 random individuals $\mathbf{P} = \{P_1, \ldots, P_5\}$ inside of the search region S and the individual database array $\mathbf{T}$, as an empty array.

Step 3:   Compute the fitness values for each individual according to the fitness calculation strategy presented in Sect. 9.3.

Step 4:   Update new evaluations in the individual database array $\mathbf{T}$.

Step 5:   Set the parameters $\rho \in [0.8, 1], \beta = 0.8, \alpha = 0.8$ and $H = 0.9$ being consistent with the gas state.

Step 6:   Apply the enhanced general procedure which is illustrated in Algorithm 9.2.

Step 7:   If the 50% of the total iteration number is completed $(1 \le k \le 0.5 \cdot gen)$, then the process continues to the liquid state procedure; otherwise go back to step 6.

Step 8:   Set the parameters $\rho \in [0.3, 0.6], \beta = 0.4, \alpha = 0.2$ and $H = 0.2$ being consistent with the liquid state.

Step 9:   Apply the enhanced general procedure which is illustrated in Algorithm 9.2.

Step 10: If the 90% (50% from the gas state and 40% from the liquid state) of the total iteration number is completed $(0.5 \cdot gen < k \le 0.9 \cdot gen)$, then the process continues to the solid state procedure; otherwise go back to step 9.

Step 11: Set the parameters $\rho \in [0.0, 0.1]$ and $\beta = 0.1, \alpha = 0$ and $H = 0$ being consistent with the solid state.

Step 12: Apply the enhanced general procedure which is illustrated in Algorithm 9.2.

Step 13:   If   the   100%   of   the   total   iteration   number   is   completed
           $(0.9 \cdot gen < k \leq gen)$, the   process   is   finished;   otherwise   go   back   to
           step 11.
Step 14:   If the number of target iterations has been reached, then determine the
           best individual (matching position) of the final population is $\hat{u}_{best}, \hat{v}_{best}$.

The proposed SMS-PD algorithm considers multiple search locations during the
complete optimization process. However, only a few of them are evaluated using
the true fitness function whereas all other remaining positions are just estimated.
Figure 9.8 shows a search-pattern that has been generated by the SMS-PD approach
considering the problem exposed in Fig. 9.7. Such pattern exhibits the evaluated
search-locations (rule 1 and 2) in red-cells, whereas the minimum location is
marked in green. Blue-cells represent those that have been estimated (rule 3)
whereas gray-intensity-cells were not visited at all, during the optimization process.

Since most of fast PD methods employ optimization algorithms that face difficulties with multi-modal surfaces, they may get trapped into local minima and find
sub-optimal detections. On the other hand, the proposed approach allows finding
out the optimal solution due to a better balance between the exploration and
exploitation of the search space. Under the effect of the SMS operators, the search
locations vary from generation to generation, avoiding to get trapped into a local
minimum. Besides, since the proposed algorithm uses a fitness calculation strategy
for reducing the evaluation of the NCC values, it requires fewer search positions.
As example Fig. 9.8 shows how the SMS-PD algorithm found the optimal detection, evaluating only the 11% of the feasible search locations.

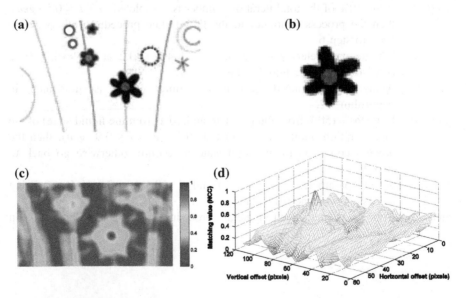

Fig. 9.7 Pattern detection process. **a** Example source image, **b** template image, **c** color-encoded
NCC values and **d** NCC multi-modal surface

Fig. 9.8 Search-pattern generated by the SMS-PD algorithm. *Red points* represent the evaluated search positions whereas *blue points* indicate the estimated locations. The *green point* exhibits the optimal match detection

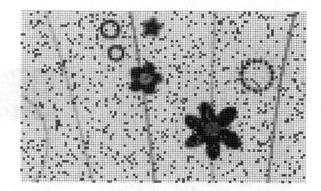

9.7 Experimental Results

In order to verify the feasibility and effectiveness of our proposed algorithm in this work, series of comparative experiments with other PD algorithms are also given. Simulations have been performed over a set of images which are shown in Fig. 9.9. The proposed approach has been applied to the experimental set whose results have been compared to those produced by the ICA-PD method [10] and the PSO-PD algorithm [8]. These are considered state-of-the-art algorithms whose results have been recently published. The maximum iteration number for the experimental set has been set to 300. Such stop criterion has been selected to maintain compatibility to similar works reported in the literature [7–10].

The parameter setting for each algorithm in the comparison is described as follows:

1. ICA-PD [10]: The parameters are set to *Num of countries* = 100, *Num of imper* = 10, *Num of colony* = 90, T_{max} = 300, ξ = 0.1, ε_1 = 0.15 and ε_2 = 0.9. Such values are the best parameter set for this algorithm according to [10].
2. PSO-PD [8]: The parameters are set to particle number = 100, c_1 = 1.5 and c_2 = 1.5; besides, the particle velocity is initialized between [−4, 4].
3. SMS-PD: The algorithm was configured by using: particle number = 100 and d = 3.

Once all algorithms were configured with such values, they are used without modification during the experiments. The comparisons are analyzed considering three performance indexes: the average elapsed time (At), the success rate (Sr), the Average number of checked locations (AsL) and the average number of function evaluations (AfE). The Average elapsed time (At) indicates the time in seconds employed during the execution of each single experiment. The success rate (Sr) represents the number of executions in percentage in which the algorithms find out successfully the optimal detection point. The Average number of checked locations (AsL) exhibits the number of search locations which has been visited during a single experiment. The average number of function evaluations

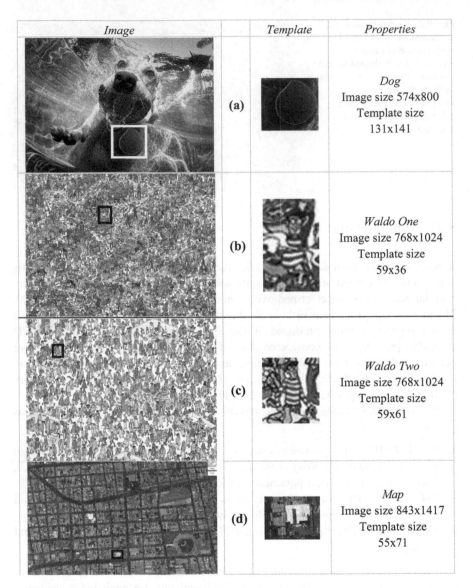

Image		Template	Properties
	(a)		*Dog* Image size 574x800 Template size 131x141
	(b)		*Waldo One* Image size 768x1024 Template size 59x36
	(c)		*Waldo Two* Image size 768x1024 Template size 59x61
	(d)		*Map* Image size 843x1417 Template size 55x71

Fig. 9.9 Experimental set used in the comparisons

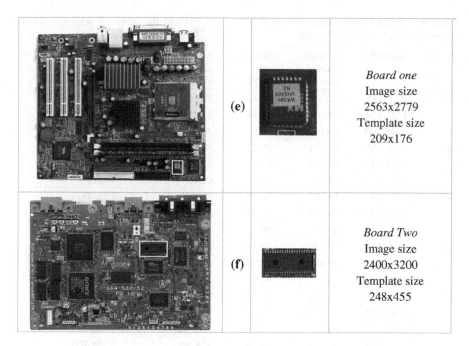

Fig. 9.9 (continued)

(AfE) indicates the number of times that the NCC coefficient is computed. In order to assure statistic consistency, all these performance indexes are averaged considering a determined number of executions.

The results for 30 runs are reported in Table 9.1 where the best outcome for each image is boldfaced. According to this table, SMS-PD delivers better results than ICA and PSO for all images. In particular, the test remarks the largest difference in the success rate (Sr) and the average number of checked locations (AsL). Such facts are directly related to a better trade-off between exploration and exploitation, and the incorporation of the fitness calculation strategy, respectively. Figure 9.10 present the matching evolution curve for each image considering the average best NCC value seen so-far for all the algorithms employed in the comparison.

From Table 9.1, it turns out that the average cost of our algorithm is 6.873 s, while the average cost of the ICA-PD and the PSO-PD algorithms are 26.331 and 25.156, respectively. Such fact demonstrates that SMS-PD spends less time on image matching in reference to its counterparts. According to Table 9.1, the SMS-PD presents a better performance than the other two algorithms in terms of effectiveness, since it detects practically in all experiments the optimal detection point. On the other hand, although the three algorithms visit approximately the same number of search location, the proposed algorithm evaluates (NCC evaluation) a minimal number of them. It is important to recall that such evaluation represents the main computational cost associated to the PD process.

Table 9.1 Performance comparison of ICA-PD, PSO-PD and the proposed approach for the experimental set shown in Fig. 9.8

Image	Algorithm	Average elapsed time (At)	Success rate (Sr)%	Average number of checked locations (AsL)	Average number of function evaluations (AfE)
(a)	ICA-TM	12.345	88.12	32,000	32,000
	PSO-TM	10.862	80.84	31,250	31,250
	SMS-TM	2.854	100	30,000	5640
(b)	ICA-TM	17.534	70.23	82,000	82,000
	PSO-TM	16.297	62.45	81,784	81,784
	SMS-TM	3.643	100	60,000	9213
(c)	ICA-TM	24.342	70.21	82,000	82,000
	PSO-TM	23.174	60.33	81,784	81,784
	SMS-TM	6.871	99	60,000	8807
(d)	ICA-TM	26.249	67.12	220,512	220,512
	PSO-TM	26.381	58.12	210,784	210,784
	SMS-TM	6.937	98	100,512	18,506
(e)	ICA-TM	37.231	90.54	578,400	578,400
	PSO-TM	35.925	85.27	578,400	578,400
	SMS-TM	10.214	100	220,512	54,341
(f)	ICA-TM	40.287	89.78	578,400	578,400
	PSO-TM	38.298	81.47	578,400	578,400
	SMS-TM	10.719	100	220,512	57,981

A non-parametric statistical significance proof known as the Wilcoxon's rank sum test for independent samples [26, 27] has been conducted over the average number of function evaluations (AfE) data of Table 9.1, with an 5% significance level. Table 9.2 reports the p values produced by Wilcoxon's test for the pair-wise comparison of the average number of function evaluations (AfE) of four groups. Such groups are formed by SMS-PD vs. ICA-PD and SMS-PD vs. PSO-PD. As a null hypothesis, it is assumed that there is no significant difference between mean values of the two algorithms. The alternative hypothesis considers a significant difference between the AfE values of both approaches.

All p values reported in Table 9.2 are less than 0.05 (5% significance level) which is a strong evidence against the null hypothesis. Therefore, such evidence indicates that SMS-PD results are statistically significant and that it has not occurred by coincidence (i.e. due to common noise contained in the process).

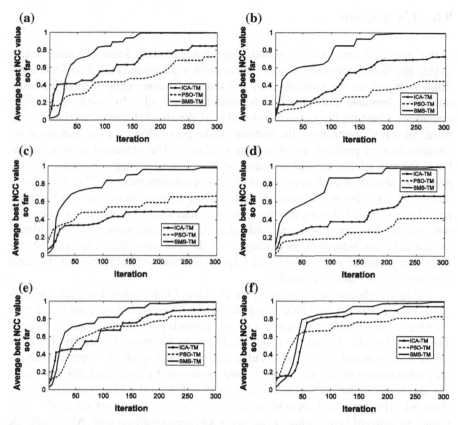

Fig. 9.10 Evolution curves for ICA-PD, PSO-PD and the proposed SMS-PD considering the average best NCC value seen so-far, each curve corresponds to the image of the experimental set

Table 9.2 p values produced by Wilcoxon's test comparing SMS-PD versus ICA-PD and SMS-PD versus PSO-PD over the average number of function evaluations (AfE) values from Table 9.1

Image	SMS-PD versus ICA-PD	SMS-PD versus PSO-PD
(a)	1.52E−10	1.78E−10
(b)	3.23E−12	5.47E−12
(c)	1.56E−12	2.67E−12
(d)	3.21E−12	5.87E−12
(e)	4.87E−12	7.58E−12
(f)	2.11E−12	4.49E−12

9.8 Conclusions

In this chapter, a novel nature-inspired algorithm, called the States of Matter Search (SMS) has been proposed for solving the pattern detection (PD). The SMS algorithm is based on the simulation of the states of matter phenomenon. In SMS, individuals emulate molecules which interact to each other by using evolutionary operations based on the physical principles of the thermal-energy motion mechanism. Such operations allow the increase of the population diversity and avoid the concentration of particles within a local minimum. The presented approach combines the use of the defined operators with a control strategy that modifies the parameter setting of each operation during the evolution process. The algorithm is devised by considering each state of matter at one different exploration–exploitation rate. Thus, the evolutionary process is divided into three stages which emulate the three states of matter: gas, liquid and solid. At each state, molecules (individuals) exhibit different behaviors. Beginning from the gas state (pure exploration), the algorithm modifies the intensities of exploration and exploitation until the solid state (pure exploitation) is reached.

The approach also incorporates a simple fitness calculation approach which is based on the Nearest Neighbor Interpolation (NNI) algorithm in order to estimate the fitness value (NCC operation) for several candidate solutions (search locations). The method is able to save computational time by identifying which NCC values can be just estimated or must be calculated instead. As a result, the approach can not only substantially reduce the number search positions (by using the SMS approach), but also to avoid the NCC evaluation for many of them (by incorporating the NNI strategy). The presented method achieves the best balance over other PD algorithms, in terms of both estimation accuracy and computational cost. As a result, the approach can substantially reduce the number of function evaluations, yet preserving the good search capabilities of SMS.

Since the proposed algorithm is designed to have a better exploration-exploitation balance than other evolutionary algorithms, a high probability for finding the true matching point (accurate detection point) is expected regardless of the high multi-modality nature of the PD process.

The performance of the proposed approach has been compared to other existing PD algorithms by considering different images which present a great variety of formats and complexities. Experimental results demonstrate the high performance of the proposed method in terms of elapsed time and the number of NCC evaluations.

References

1. Brunelli R (2009) Template matching techniques in computer vision: theory and practice. Wiley. ISBN: 978-0-470-51706-2
2. Hadi G, Mojtaba L, Hadi SY (2009) An improved pattern matching technique for lossy/lossless compression of binary printed Farsi and Arabic textual images. Int J Intell Comput Cybern 2(1):120–147
3. Krattenthaler W, Mayer KJ, Zeiler M (1994) Point correlation: a reduced-cost template matching technique. In: Proceedings of the first IEEE international conference on image processing, pp 208–212
4. Rosenfeld A, VanderBrug GJ (1977) Coarse-fine template matching. IEEE Trans Syst Man Cybern, SMC-7(2):104–107
5. Tanimoto SL (1981) Template matching in pyramids. Comput Graph Image Process 16 (4):356–369
6. Uenohara M, Kanade T (1997) Use of Fourier and Karhunen-Loeve decomposition for fast pattern matching with a large set of templates. IEEE Trans Pattern Anal Mach Intell 19 (8):891–898
7. Dong N, Wu C-H, Ip W-H, Chen Z-Q, Chan C-Y, Yung K-L (2011) An improved species based genetic algorithm and its application in multiple template matching for embroidered pattern inspection. Expert Syst Appl 38:15172–15182
8. Liu F, Duana H, Deng Y (2012) A chaotic quantum-behaved particle swarm optimization based on lateral inhibition for image matching. Optik 123:1955–1960
9. Wu C-H, Wang D-Z, Ip A, Wang D-W, Chan C-Y, Wang H-F (2009) A particle swarm optimization approach for components placement inspection on printed circuit boards. J Intell Manuf 20:535–549
10. Duan H, Xu C, Liu S, Shao S (2010) Template matching using chaotic imperialist competitive algorithm. Pattern Recogn Lett 31:1868–1875
11. Chen G, Low CP, Yang Z (2009) Preserving and exploiting genetic diversity in evolutionary programming algorithms. IEEE Trans Evol Comput 13(3):661–673
12. Adra SF, Fleming PJ (2011) Diversity management in evolutionary many-objective optimization. IEEE Trans Evol Comput 15(2):183–195
13. Tan KC, Chiam SC, Mamun AA, Goh CK (2009) Balancing exploration and exploitation with adaptive variation for evolutionary multi-objective optimization. Eur J Oper Res 197:701–713
14. Jin Y (2005) Comprehensive survey of fitness approximation in evolutionary computation. Soft Comput 9:3–12
15. Jin Yaochu (2011) Surrogate-assisted evolutionary computation: recent advances and future challenges. Swarm Evol Comput 1:61–70
16. Branke J, Schmidt C (2005) Faster convergence by means of fitness estimation. Soft Comput 9:13–20
17. Zhou Z, Ong Y, Nguyen M, Lim D (2005) A study on polynomial regression and gaussian process global surrogate model in hierarchical surrogate-assisted evolutionary algorithm. IEEE congress on evolutionary computation (ECiDUE'05), Edinburgh, United Kingdom, 2–5 Sept 2005
18. Ratle A (2001) Kriging as a surrogate fitness landscape in evolutionary optimization. Artif Intell Eng Des Anal Manuf 15:37–49
19. Lim D, Jin Y, Ong Y, Sendhoff B (2010) Generalizing surrogate-assisted evolutionary computation. IEEE Trans Evol Comput 14(3):329–355
20. Ong Y, Lum K, Nair P (2008) Evolutionary algorithm with hermite radial basis function interpolants for computationally expensive adjoint solvers. Comput Optim Appl 39(1):97–119
21. Ceruti G, Rubin H (2007) Infodynamics: analogical analysis of states of matter and information. Inf Sci 177:969–987

22. Debashish C, Dietrich S (2000) Principles of equilibrium statistical mechanics, 1° edn. Wiley-VCH
23. Yunus AC, Boles MA (2005) Thermodynamics: an engineering approach, 5 edn. McGraw-Hill
24. Frederick B, Eugene H. Schaum's outline of college physics, 11th edn. McGraw-Hill
25. David SB, Roy E (1992) Turner Introductory statistical mechanics, 1° edn. Addison Wesley
26. Wilcoxon F (1945) Individual comparisons by ranking methods. Biometrics 1:80–83
27. Garcia S, Molina D, Lozano M, Herrera F (2008) A study on the use of non-parametric tests for analyzing the evolutionary algorithms' behaviour: a case study on the CEC'2005 Special session on real parameter optimization. J Heurist. doi:10.1007/s10732-008-9080-4

Chapter 10
Artificial Bee Colony Algorithm Applied to Multi-threshold Segmentation

Image segmentation is a very important task in Computer Vision community, due to its capabilities for further steps that lead to recognizing patterns in digital images. Thus, the process of thresholding selection has become an interesting area, in recent years this procedure has been investigated as an optimization problem. On the other Hand, ABC is a nature inspired algorithm based on the intelligent behaviour of honey-bees which has been successfully used to solve complex real life optimization problems. In this chapter, a multi-thresholding approach, in which an image 1-D histogram is approximated by means of a Gaussian mixture model is presented. In the approach, the parameters are calculated by using the ABC algorithm. Under this method, each Gaussian function represents a pixel class; hence a threshold. The presented ABC approach shows fast convergence and low sensitivity to initial conditions. Experimental results has shown the ABC-method's capability to perform multi-threshold selection and interesting advantages in comparison to other algorithms.

10.1 Introduction

Several image processing applications aim to detect and classify relevant features which may be later analyzed to perform several high-level tasks. In particular, image segmentation seeks to group pixels within meaningful regions. Commonly, gray levels belonging to the object, are substantially different from those featuring the background. Thresholding is thus a simple but effective tool to isolate objects of interest; its applications include several classics such as document image analysis, whose goal is to extract printed characters [1, 2], logos, graphical content, or musical scores; also it is used for map processing which aims to locate lines, legends, and characters [3]. Moreover, it is employed for scene processing, seeking for object detection, marking [4] and for quality inspection of materials [5, 6].

© Springer International Publishing AG 2017
M.-A. Díaz-Cortés et al., *Engineering Applications of Soft Computing*,
Intelligent Systems Reference Library 129, DOI 10.1007/978-3-319-57813-2_10

Thresholding selection techniques can be classified into two categories: bi-level and multi-level. In the former, one limit value is chosen to segment an image into two classes: one representing the object and the other one segmenting the background. When distinct objects are depicted within a given scene, multiple threshold values have to be selected for proper segmentation, which is commonly called multilevel thresholding.

A variety of thresholding approaches have been proposed for image segmentation, including conventional methods [7–10] and intelligent techniques [11, 12]. Extending the segmentation algorithms to a multilevel approach may cause some inconveniences: (1) they may have no systematic or analytic solution when the number of classes to be detected increases and (2) they may also show a slow convergence and/or high computational cost [11].

In this work, the segmentation algorithm is based on a parametric model holding a probability density function of gray levels which groups a mixture of several Gaussian density functions (Gaussian mixture). Mixtures represent a flexible method of statistical modelling as they are employed in a wide variety of contexts [13]. Gaussian mixture has received considerable attention in the development of segmentation algorithms despite its performance is influenced by the shape of the image histogram and the accuracy of the estimated model parameters [14]. The associated parameters can be calculated considering the expectation maximization (EM) algorithm [15, 16] or Gradient-based methods such as Levenberg–Marquardt, LM [17]. However, EM algorithms are very sensitive to the choice of the initial values [18], meanwhile Gradient-based methods are computationally expensive and may easily get stuck within local minima [14]. Therefore, some researchers have attempted to develop methods based on modern global optimization algorithms such as the Learning Automata (LA) [19] and differential evolution algorithm [20]. In this chapter, an alternative approach using an optimization algorithm for determining the parameters of a Gaussian mixture is presented.

On other hand, Karaboga [21] has presented a metaheuristic algorithm for solving numerical optimization problems known as the artificial bee colony (ABC) method. Inspired by the intelligent foraging behavior of a honeybee swarm, the ABC algorithm consists of three essential components: food source positions, nectar-amounts and several honey-bee classes. Each food source position represents a feasible solution for the problem under consideration. The nectar-amount for a food source represents the quality of such solution according to its fitness value. Each bee-class symbolizes one particular operation for generating new candidate food source positions (i.e. candidate solutions).

The ABC algorithm starts by producing a randomly distributed initial population (food source locations). After initialization, an objective function evaluates whether such candidates represent an acceptable solution (nectar-amount) or not. Guided by the values of such objective function, candidate solutions are evolved through different ABC operations (honey-bee types). When the fitness function (nectar-amount) cannot be further improved after a maximum number of cycles, its related food source is assumed to be abandoned and replaced by a new randomly chosen food source location.

The performance of ABC algorithm has been compared to other metaheuristic methods such as genetic algorithms (GA), differential evolution (DE) and particle swarm optimization (PSO) [22, 23]. The results have shown that ABC can produce optimal solutions yet more effectively than other methods for several optimization problems. Such characteristics have motivated the use of ABC to solve different sorts of engineering problems within different fields such as signal processing [24], flow shop scheduling [25], structural inverse analysis [26], clustering [27, 28] and electromagnetism [29].

This chapter presents the use of the Artificial Bee Colony (ABC) algorithm to compute threshold selection for image segmentation. In this approach, the segmentation process is considered as an optimization problem approximating the 1-D histogram of a given image by means of a Gaussian mixture model. The operation parameters are calculated through the ABC algorithm. Each Gaussian function approximating the histogram represents a pixel class and therefore a threshold point in the segmentation scheme. The experimental results, presented in this work, demonstrate that ABC exhibits fast convergence, relative low computational cost and no sensitivity to initial conditions by keeping an acceptable segmentation of the image, i.e. a better mixture approximation in comparison to the EM or gradient based algorithms.

The chapter is organized as follows: Sect. 10.2 presents the Gaussian approximation of the histogram while Sect. 10.3 discusses on the ABC algorithm. Section 10.4 formulates the threshold determination with Sect. 10.5 presenting all experimental results after the presented approach is implemented. Section 10.6 summarizes a full discussion on the algorithm performance.

10.2 Gaussian Approximation

Let consider an image holding L gray levels $[0, \ldots, L - 1]$ whose distribution is displayed within a histogram $h(g)$. In order to simplify the description, the histogram is normalized just as a probability distribution function, yielding:

$$h(g) = \frac{n_g}{N}, \ h(g) > 0,$$

$$N = \sum_{g=0}^{L-1} n_g, \ \text{and} \ \sum_{g=0}^{L-1} h(g) = 1, \quad (10.1)$$

where n_g denotes the number of pixels with gray level g and N being the total number of pixels in the image. The histogram function can thus be contained into a mix of Gaussian probability functions of the form:

$$p(x) = \sum_{i=1}^{K} P_i \cdot p_i(x) = \sum_{i=1}^{K} \frac{P_i}{\sqrt{2\pi}\sigma_i} \exp\left[\frac{-(x - \mu_i)^2}{2\sigma_i^2}\right] \qquad (10.2)$$

with P_i being the probability of class i, $p_i(x)$ being the probability distribution function of gray-level random variable x in class i, with μ_i and σ_i being the mean and standard deviation of the i-th probability distribution function and K being the number of classes within the image. In addition, the constraint $\sum_{i=1}^{K} P_i = 1$ must be satisfied.

The mean square error is used to estimate the $3K$ parameters P_i, μ_i and σ_i, $i = 1$, ..., K. For instance, the mean square error between the Gaussian mixture $p(x_i)$ and the experimental histogram function $h(x_i)$ is defined as follows:

$$J = \frac{1}{n} \sum_{j=1}^{n} \left[p(x_j) - h(x_j)\right]^2 + \omega \cdot \left|\left(\sum_{i=1}^{K} P_i\right) - 1\right| \qquad (10.3)$$

Assuming an n-point histogram as in [13] and ω being the penalty associated with the constrain $\sum_{i=1}^{K} P_i = 1$.

In general, the parameter estimation that minimizes the square error produced by the Gaussian mixture is not a simple problem. A straightforward method is to consider the partial derivatives of the error function to zero by obtaining a set of simultaneous transcendental equations [13]. However, an analytical solution is not always available considering the non-linear nature of the equation which in turn yields the use of iterative approaches such as gradient-based or maximum likelihood estimation. Unfortunately, such methods may also get easily stuck within local minima or be time expensive.

In the case of other algorithms such as the EM algorithm and the gradient-based methods, the new parameter point lies within a neighbourhood distance of the previous parameter point. However, this is not the case for the ABC adaptation algorithm which is based on stochastic principles. New operating points are thus determined by a parameter probability function that may yield points lying far away from previous operating points, providing the algorithm with a higher ability to locate and pursue a global minimum.

10.3 Artificial Bee Colony (ABC) Algorithm

The ABC algorithm assumes the existence of a set of operations that may resemble some features of the honey bee behavior. For instance, each solution within the search space includes a parameter set representing food source locations. The "fitness value" refers to the food source quality that is strongly linked to the food's location. The process mimics the bee's search for valuable food sources yielding an analogous process for finding the optimal solution.

10.3.1 Biological Bee Profile

The minimal model for a honey bee colony consists of three classes: employed bees, onlooker bees and scout bees. The employed bees will be responsible for investigating the food sources and sharing the information with recruit onlooker bees. They, in turn, will make a decision on choosing food sources by considering such information. The food source having a higher quality will have a larger chance to be selected by onlooker bees than those showing a lower quality. An employed bee, whose food source is rejected as low quality by employed and onlooker bees, will change to a scout bee to randomly search for new food sources. Therefore, the exploitation is driven by employed and onlooker bees while the exploration is maintained by scout bees. The implementation details of such bee-like operations in the ABC algorithm are described in the next sub-section.

10.3.2 Description of the ABC Algorithm

Resembling other metaheuristic approaches, the ABC algorithm is an iterative process. It starts with a population of randomly generated solutions or food sources. The following three operations are applied until a termination criterion is met [23]:

1. Send the employed bees.
2. Select the food sources by the onlooker bees.
3. Determine the scout bees.

10.3.2.1 Initializing the Population

The algorithm begins by initializing N_p food sources. Each food source is a D-dimensional vector containing the parameter values to be optimized, which are randomly and uniformly distributed between the pre-specified lower initial parameter bound x_j^{low} and the upper initial parameter bound x_j^{high}.

$$x_{j,i} = x_j^{low} + \text{rand}(0, 1) \cdot (x_j^{high} - x_j^{low});$$
$$j = 1, 2, \ldots, D; \ i = 1, 2, \ldots, N_p. \tag{10.4}$$

with j and i being the parameter and individual indexes respectively. Hence, $x_{j,i}$ is the jth parameter of the ith individual.

10.3.2.2   Send Employed Bees

The number of employed bees is equal to the number of food sources. At this stage, each employed bee generates a new food source in the neighborhood of its present position as follows:

$$
\begin{aligned}
v_{j,i} &= x_{j,i} + \phi_{j,i}(x_{j,i} - x_{j,k}); \\
k &\in \{1, 2, \ldots, N_p\}; j \in \{1, 2, \ldots, D\}
\end{aligned}
\tag{10.5}
$$

$x_{j,i}$ is a randomly chosen j parameter of the ith individual and k is one of the N_p food sources, satisfying the condition $i \neq k$. If a given parameter of the candidate solution v_i exceeds its predetermined boundaries, that parameter should be adjusted in order to fit the appropriate range. The scale factor $\phi_{j,i}$ is a random number between $[-1, 1]$. Once a new solution is generated, a fitness value representing the profitability associated with a particular solution is calculated. The fitness value for a minimization problem can be assigned to each solution v_i by the following expression:

$$
fit_i = \begin{cases} \frac{1}{1+J_i} & \text{if } J_i \geq 0 \\ 1 + abs(J_i) & \text{if } J_i < 0 \end{cases}
\tag{10.6}
$$

where J_i is the objective function to be minimized. A greedy selection process is thus applied between v_i and x_i. If the nectar- amount (fitness) of v_i is better, then the solution x_i is replaced by v_i; otherwise, x_i remains.

10.3.2.3   Select the Food Sources by the Onlooker Bees

Each onlooker bee (the number of onlooker bees corresponds to the food source number) selects one of the proposed food sources, depending on their fitness value, which has been recently defined by the employed bees. The probability that a food source will be selected can be obtained from the following equation:

$$
Prob_i = \frac{fit_i}{\sum_{i=1}^{N_p} fit_i}
\tag{10.7}
$$

where fit_i is the fitness value of the food source i, which is related to the objective function value (J_i) corresponding to the food source i. The probability of a food source being selected by onlooker bees increases with an increase in the fitness value of the food source. After the food source is selected, onlooker bees will go to the selected food source and select a new candidate food source position inside the neighborhood of the selected food source. The new candidate food source can be expressed and calculated by Eq. (10.5). In case the nectar-amount, i.e., fitness of the new solution, is better than before, such position is held; otherwise, the last solution remains.

10.3.2.4 Determine the Scout Bees

If a food source i (candidate solution) cannot be further improved through a pre-determined trial number known as "limit", the food source is assumed to be abandoned and the corresponding employed or onlooker bee becomes a scout. A scout bee explores the searching space with no previous information, i.e., the new solution is generated randomly as indicated by Eq. (10.4). In order to verify if a candidate solution has reached the predetermined "*limit*", a counter A_i is assigned to each food source i. Such a counter is incremented consequent to a bee-operation failing to improve the food source's fitness.

10.4 Determination of Thresholding Values

In order to determine optimal threshold values, it is considered that the data classes are organized such that $\mu_1 < \mu_2 < \cdots < \mu_K$. Therefore, threshold values are obtained by computing the overall probability error of two adjacent Gaussian functions, yielding:

$$E(T_h) = P_{h+1} \cdot E_1(T_h) + P_i \cdot E_2(T_h),$$
$$h = 1, 2, \ldots, K - 1 \tag{10.8}$$

considering

$$E_1(T_h) = \int_{-\infty}^{T_h} p_{h+1}(x)dx, \tag{10.9}$$

and

$$E_2(T_h) = \int_{T_h}^{\infty} p_h(x)dx, \tag{10.10}$$

$E_1(T_h)$ is the probability of mistakenly classifying the pixels in the $(h + 1)$-th class belonging to the h-th class, while $E_2(T_h)$ is the probability of erroneously classifying the pixels in the h-th class belonging to the $(h + 1)$-th class. P'_js are the a priori probabilities within the combined probability density function, and T_h is the threshold value between the h-th and the $(h + 1)$-th classes. One T_h value is chosen such as the error $E(T_h)$ is minimized. By differentiating $E(T_h)$ with respect to T_h and equating the result to zero, it is possible to use the following equation to define the optimum threshold value T_h:

$$AT_h^2 + BT_h + C = 0 \tag{10.11}$$

considering

$$
\begin{aligned}
A &= \sigma_h^2 - \sigma_{h+1}^2 \\
B &= 2 \cdot (\mu_h \sigma_{h+1}^2 - \mu_{h+1} \sigma_h^2) \\
C &= (\sigma_h \mu_{h+1})^2 - (\sigma_{h+1} \mu_h)^2 + 2 \cdot (\sigma_h \sigma_{h+1})^2 \cdot \ln\left(\frac{\sigma_{h+1} P_h}{\sigma_h P_{h+1}}\right)
\end{aligned}
\tag{10.12}
$$

Although the above quadratic equation has two possible solutions, only one of them is feasible, i.e. a positive value falling within the interval.

10.5 Experimental Results

By considering that the mixture parameters are extracted from the fitness function J (Eq. 10.3) after applying the ABC algorithm, three experiments are set to evaluate the performance of the presented algorithm. The first one considers the well-known image "The Camera-man" which is shown by Fig. 10.1a, with its corresponding histogram shown by Fig. 10.1b. The goal is to segment the image in three different pixel classes. According to Eq. (10.2), during learning, the ABC algorithm adjusts nine parameters, following the minimization procedure conducted by Eq. (10.3). In this experiment, a population of 20 (N_p) bees is considered, with ten employed and ten onlookers bees. Each candidate holds 9 dimensions, such as:

$$I_N = \left\{ P_1^N, \sigma_1^N, \mu_1^N, P_2^N, \sigma_2^N, \mu_2^N, P_3^N, \sigma_3^N, \mu_3^N \right\} \tag{10.13}$$

with N representing the individual's number. The parameters (P, σ, μ) are randomly initialized, but assuming some restrictions to each parameter (for example μ must fall between 0 and 255).

The experiments suggest that after 200 iterations, the ABC algorithm has converged to the global minimum. Figure 10.2a shows the obtained Gaussian functions (pixel classes) plotted over the original histogram while Fig. 10.2b shows the Gaussian mixture. Figure 10.3 shows the segmented image whose threshold values are calculated according to Eqs. (10.11) and (10.12).

The algorithm is tested with a greater number of Gaussian functions yielding the need of optimizing more parameters (according to Eq. 10.3). Thus, twelve parameters are now considered corresponding to the values of four Gaussian functions. One population of 20 bees and 12 dimensions are used for the test.

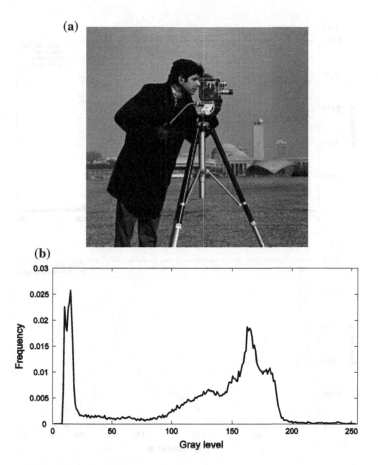

Fig. 10.1 **a** Original image "The Cameraman" and **b** its correspondent histogram

Figure 10.4a shows the Gaussian functions (pixel classes) plotted over the histogram after 200 iterations, while in Fig. 10.4b presents the Gaussian mixture. The segmented image is depicted by Fig. 10.5.

The second experiment considers the popular benchmark image known as "The scene" (see Fig. 10.6a). The image's histogram is presented by Fig. 10.6b. Following the first experiment, the image is segmented considering four pixel classes. The optimization is performed by the ABC algorithm which results in the classes shown by Fig. 10.7a. In turn, Fig. 10.7b presents the Gaussian mixture as it results from the addition of other Gaussian functions. Figure 10.8 shows the image segmentation considering four classes.

The final experiment considers a blood-smear image as it is processed by the ABC algorithm. The image shows a set of leukocytes cells in the blood smear

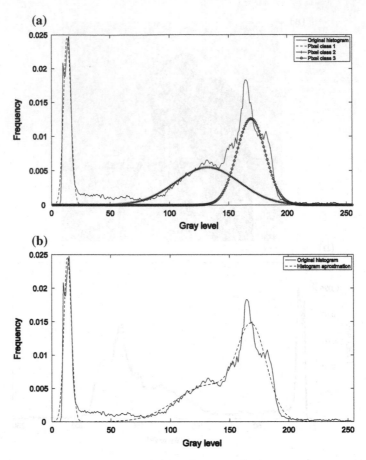

Fig. 10.2 Applying the ABC algorithm for 3 classes and its results: **a** Gaussian functions for each class and **b** mixed Gaussian functions (approach to the original histogram)

(darker cells on image of Fig. 10.9a). The gray blobs represent the red blood cells while the background is white yielding only three classes to be considered.

10.5.1 Comparing the ABC Algorithm Versus the EM and LM Methods

This section discusses on the comparison between ABC and other algorithms such as the EM algorithm and the Levenberg–Marquardt (LM) method which are commonly employed for determining Gaussian mixtures. The discussion focuses on

Fig. 10.3 Segmented image
considering only three classes

the following issues: sensitivity to the initial conditions, convergence and computational costs.

(a) Sensitivity to initial conditions. This experiment considers different initial values for all methods assuming the same histogram in the approximation task. After convergence, only final parameters representing the Gaussian mixture are reported. Figure 10.10a shows the image used in the comparison while Fig. 10.10b pictures the histogram. All experiments are conducted several times in order to assure consistency. Only two different initial states with the highest variation are reported in Table 10.1. Likewise, Fig. 10.11 shows the obtained segmented images considering two initial conditions as it is reported by Table 10.1. In the ABC case, the algorithm does not require initialization as random initial values are employed. However, in order to assure a valid comparison, same initial values are considered for the EM, the LM and the ABC method.

By analyzing the information in Table 10.1, the sensitivity of the EM algorithm to initial conditions becomes evident. Figure 10.11 shows a clear pixel misclassification in some sections of the image as a consequence of such sensitivity.

(b) Convergence and computational cost. The experiment aims to measure the number of required steps and the computing time spent by the EM, the LM and the ABC algorithm required to calculate the parameters of the Gaussian mixture in benchmark images (see Fig. 10.12a–c). All experiments consider four classes. Table 10.2 shows the averaged measurements as they are obtained from 20 experiments. It is evident that the EM is the slowest to converge (iterations) and the LM shows the highest computational cost (time elapsed) because it requires complex Hessian approximations. On the other hand, the ABC shows an acceptable trade off between its convergence time and its computational cost. Finally, Fig. 10.12 below shows the segmented images as they are generated by each algorithm.

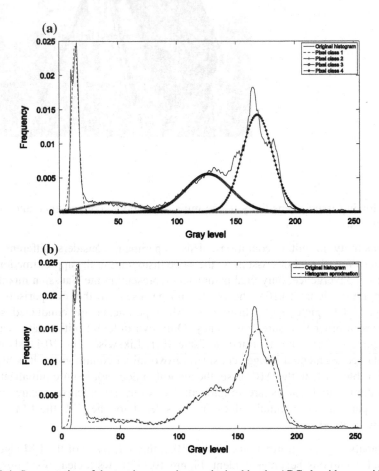

Fig. 10.4 Segmentation of the test image as it was obtained by the ABC algorithm considering 4 classes: **a** Gaussian functions for each class and **b** mixed Gaussian functions approaching the original histogram

Fig. 10.5 Segmented image
considering four classes

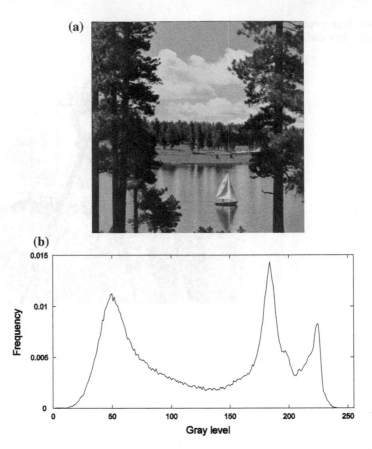

Fig. 10.6 Second experiment, **a** the original image "The scene" and **b** its histogram

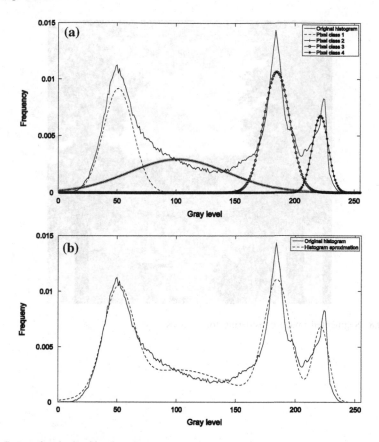

Fig. 10.7 Results obtained by the ABC algorithm for 4 classes: **a** Gaussian functions at each class and **b** mixed Gaussian functions approaching the original histogram

Fig. 10.8 Segmented image considering four classes

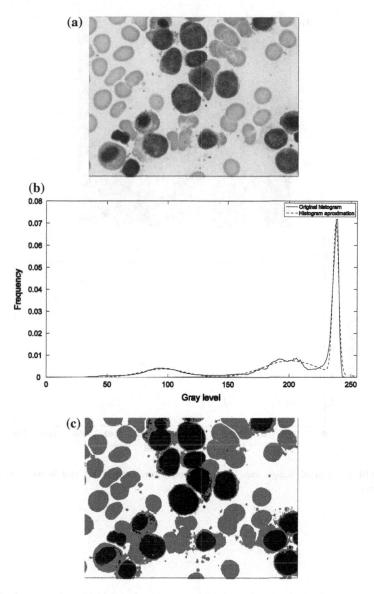

Fig. 10.9 Segmentation of a blood smear image considering three classes for the ABC algorithm: **a** original image, **b** comparison between the original histogram and the Gaussian approach, **c** the segmented image

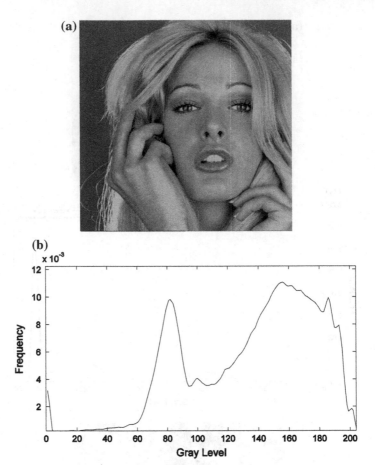

Fig. 10.10 **a** Original image used for the comparison experiment and **b** its corresponding histogram

Table 10.1 Comparison between the EM, the LM and the ABC algorithm, considering two different initial conditions

Parameters	Initial condition 1	EM	LM	ABC	Initial condition 2	EM	LM	ABC
μ_1	40.6	33.13	32.12	32.01	10	20.90	31.80	32.50
μ_2	81.2	81.02	82.05	82.30	100	82.78	80.85	82.42
μ_3	121.8	127.52	127	127.00	138	146.67	128	127.72
μ_4	162.4	167.58	166.80	166.10	200	180.72	165.90	166.50
σ_1	15	25.90	25.50	25.30	10	18.52	20.10	25.01
σ_2	15	9.78	9.70	9.80	5	12.52	9.81	10.00
σ_3	15	17.72	17.05	17.71	8	20.5	15.15	17.57
σ_4	15	17.03	17.52	17.21	22	10.09	18.00	17.22
P_1	0.25	0.0313	0.0310	0.307	0.20	0.0225	0.0312	0.317
P_2	0.25	0.2078	0.2081	0.201	0.30	0.2446	0.2079	0.255
P_3	0.25	0.2508	0.2500	0.249	0.20	0.5232	0.2502	0.260
P_4	0.25	0.5102	0.5110	0.555	0.30	0.2098	0.5108	0.511

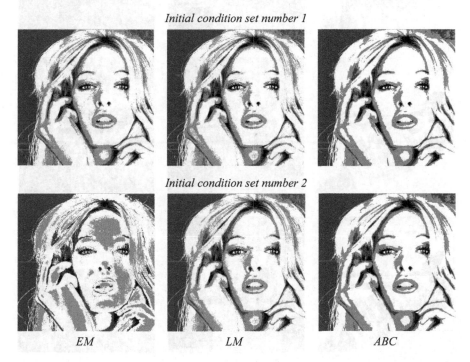

Initial condition set number 1

Initial condition set number 2

EM   LM   ABC

Fig. 10.11 Segmented images after applying the EM, the LM and the ABC algorithm with different initial conditions

(a)   **(b)**   **(c)**

Original images

EM segmented images

LM segmented images

ABC segmented images

Fig. 10.12 Original benchmark images (**a–c**), and segmented images obtained by the EM, the LM and the ABC algorithms

Table 10.2 Iterations and time requirements of the EM, the LM and the ABC algorithm as they are applied to segment benchmark images (see Fig. 10.12)

Iterations Time elapsed	a	b	c
EM	1855	1833	1870
	2.72 s	2.70 s	2.73 s
LM	985	988	958
	4.03 s	4.04 s	4.98 s
ABC	409	399	512
	0.78 s	0.70 s	0.81 s

10.6 Conclusions

In this chapter, an automatic image multi-threshold approach based on the Artificial Bee Colony (ABC) algorithm is presented. The segmentation process is considered to be similar to an optimization problem. The algorithm approximates the 1-D histogram of a given image using a Gaussian mixture model whose parameters are calculated through the ABC algorithm. Each Gaussian function approximating the histogram represents a pixel class and therefore one threshold point.

Experimental evidence shows that the ABC algorithm has an acceptable compromise between its convergence time and its computational cost when it is compared to the expectation-maximization (EM) method and the Levenberg–Marquardt (LM) algorithm. Additionally, the ABC algorithm also exhibits a better performance under certain circumstances (initial conditions) on which it is well-reported in the literature [18] that the EM has underperformed. Finally, the results have shown that the stochastic search accomplished by the ABC method shows a consistent performance with no regard of the initial value and still showing a greater chance to reach the global minimum.

References

1. Abak T, Baris U, Sankur B (1997) The performance of thresholding algorithms for optical character recognition. In: Proceedings of international conference on document analytical recognition, pp 697–700
2. Kamel M, Zhao A (1993) Extraction of binary character/graphics images from grayscale document images, Graph. Models Image Process 55(3):203–217
3. Trier OD, Jain AK (1995) Goal-directed evaluation of binarization methods. IEEE Trans Pattern Anal Mach Intell 17(12):1191–1201
4. Bhanu B (1986) Automatic target recognition: state of the art survey. IEEE Trans Aerosp Electron Syst 22:364–379
5. Sezgin M, Sankur B (2001) Comparison of thresholding methods for non-destructive testing applications, in: IEEE international conference on image processing, pp 764–767

6. Sezgin M, Tasaltin R (2000) A new dichotomization technique to multilevel thresholding devoted to inspection applications. Pattern Recognit Lett 21(2):151–161
7. Guo R, Pandit SM (1998) Automatic threshold selection based on histogram modes and discriminant criterion. Mach Vis Appl 10:331–338
8. Pal NR, Pal SK (1993) A review on image segmentation techniques. Pattern Recognit 26:1277–1294
9. Shaoo PK, Soltani S, Wong AKC, Chen YC (1988) Survey: a survey of thresholding techniques. Comput Vis Graph Image Process 41:233–260
10. Snyder W, Bilbro G, Logenthiran A, Rajala S (1990) Optimal thresholding: a new approach. Pattern Recognit Lett 11:803–810
11. Chen S, Wang M (2005) Seeking multi-thresholds directly from support vectors for image segmentation. Neurocomputing 67(4):335–344
12. Chih-Chih L (2006) A novel image segmentation approach based on particle swarm optimization. IEICE Trans Fundam 89(1):324–327
13. Gonzalez RC, Woods RE (1992) Digital image processing. Addison Wesley, Reading
14. Gupta L, Sortrakul T (1998) A Gaussian-mixture-based image segmentation algorithm. Pattern Recognit 31(3):315–325
15. Dempster AP, Laird AP, Rubin DB (1977) Maximum likelihood from incomplete data via the EM algorithm. J R Stat Soc Ser B 39(1):1–38
16. Zhang Z, Chen C, Sun J, Chan L (2003) EM algorithms for Gaussian mixtures with split-and-merge operation. Pattern Recognit 36:1973–1983
17. Park H, Amari S, Fukumizu K (2000) Adaptive natural gradient learning algorithms for various stochastic models. Neural Netw 13:755–764
18. Park H, Ozeki T (2009) Singularity and slow convergence of the EM algorithm for Gaussian mixtures. Neural Process Lett 29:45–59
19. Cuevas E, Zaldivar D, Perez-Cisneros M (2010) Seeking multi-thresholds for image segmentation with Learning Automata. Mach Vis Appl. doi:10.1007/s00138-010-0249-0
20. Cuevas E, Zaldivar D, Perez-Cisneros M (2010) A novel multi-threshold segmentation approach based on differential evolution optimization. Expert Syst With Appl 37(7):5265–5271
21. Karaboga D (2005) An idea based on honey bee swarm for numerical optimization, technical report-TR06. Erciyes University, Engineering Faculty, Computer Engineering Department
22. Karaboga D, Basturk B (2008) On the performance of artificial bee colony (ABC) algorithm. Appl Soft Comput 8(1):687–697
23. Karaboga D, Akay B (2009) A comparative study of artificial bee colony algorithm. Appl Math Comput 214:108–132
24. Karaboga N (2009) A new design method based on artificial bee colony algorithm for digital IIR filters. J Franklin Inst 346:328–348
25. Pan QK, Tasgetiren MF, Suganthan PN, Chua TJ. A discrete artificial bee colony algorithm for the lot-streaming flow shop scheduling problem. Inform Sci. doi:10.1016/j.ins.2009.12.025
26. Kang Fei, Li Junjie, Qing Xu (2009) Structural inverse analysis by hybrid simplex artificial bee colony algorithms. Comput Struct 87:861–870
27. Zhang Changsheng, Ouyang Dantong, Ning Jiaxu (2010) An artificial bee colony approach for clustering. Expert Syst Appl 37:4761–4767
28. Karaboga Dervis, Ozturk Celal (2011) A novel clustering approach: artificial bee colony (ABC) algorithm. Appl Soft Comput 11:652–657
29. Ho SL, Yang S (2009) An artificial bee colony algorithm for inverse problems. Int J Appl Electromagn Mech 31:181–192

Chapter 11
Learning Automata Applied to Planning Control

Planning Control uses information regarding a problem and its environment to decide whether one plan is better than other in order to reach a required control objective. An interesting alternative for planning control is model predictive control (MPC) and the receding horizon control. MPC is the planning approach that has recently found a wide acceptance for industrial applications. This chapter explores the usefulness of planning to improve the performance of feedback-based control schemes considering a probabilistic approach known as Learning Automata (LA). The LA algorithms are based on stochastic principles, where new search positions are determined by a probability function without considering how close they are from previous existent solutions. The presented LA approach may be considered as a planning system that chooses the strategy with the higher probability to produce the best closed-loop performance results. The effectiveness of the methodology is tested over a nonlinear plant and compared with the results offered by the Levenberg–Marquardt (LM) algorithm.

11.1 Introduction

Planning requires the ability to build representations like all the daily-life models stored in our brain. In turn, this fact allows us to generate predictions on how the environment would react to several plans. The ability of choosing among different alternative plans and executing among several sequences of actions, it is exclusively performed by humans. Planning is the approach which allows generating complex behavior surpassing the simple reaction to what is sensed. Planning Control uses information about the problem and its environment, often embedded into some type of a model, to consider many options (also known as plans), choosing the best one to achieve the required objectives in the control loop.

Planning also provides a very general and easy methodology to apply. It has been exploited extensively in conventional control, e.g. receding horizon control

© Springer International Publishing AG 2017
M.-A. Díaz-Cortés et al., *Engineering Applications of Soft Computing*,
Intelligent Systems Reference Library 129, DOI 10.1007/978-3-319-57813-2_11

and model predictive control. In comparison to intelligent approaches as neural networks [1] or Evolutionary algorithms [2], it exploits the use of an explicit approximated model to decide what actions to take. However, like the fuzzy and expert system approaches, it is still possible to incorporate heuristics that help to specify which control actions are the best to use. In broad sense, planning approaches attempt to use both heuristic knowledge and model-based knowledge in order to make control decisions. This may be the fundamental reason for selecting a planning strategy over a simple rule-based system. It is a bad engineering practice to prefer the use of heuristics and ignore the information provided by a good mathematical model while planning strategies provide a way to incorporate this information.

Planning has been successfully used in the solution of several engineering problems [3, 4]. However, in the case of controlling dynamic systems, publications are quite few some examples are given in [5, 6]. There exist different planning system approaches depending on the type of problems and the quantity of plans that is considered for the solution. An example of these approaches is Belief-Desire-Intention [7], which is effective in those cases in which the plan number is finite and sensibly small, reason why it is little used in control processes.

Model Predictive Control [8] is the planning approach that has recently found a wide acceptance for industrial applications. The generation of control signals in MPC involves the on-line use of one parametric model of the plant, assuming an efficient control plan. Major design techniques of MPC include Model Algorithm Control, Dynamic Matrix Control, Internal Model Control and Generalized Predictive Control, among others [9]. The strategy of MPC is, at any given time, to solve on-the-fly a receding open-loop optimal control problem over a finite time horizon, taking only the first result in the control sequence. MPC algorithms are very intuitive and easy to understand with practical constraints commonly included in the on-line algorithm [10]. MPC has received worldwide attention because it is straightforward to implement in industrial applications, particularly in chemical processes, where dynamics are relatively slow, easily accommodating the on-line optimization [9].

Several variants of the MPC methodology have appeared in the literature. In [11], the plan evaluation is done over non-linear models including an industrial batch reactor in [12]. Furthermore, the idea of mixing iterative learning control to feedback model based control is presented in [13]. Recently, it has been proposed the use of Set Membership (SM) methodologies in the approximation of Model Predictive Control (MPC) and the laws for linear systems in [14]. However, much of the work has been limited to an optimization strategy (plan evaluation and selection) based on dynamic programming and gradient methods. The use of these techniques optimization, considering most of the non-linear control problems, is multimodal yielding a slow speed operation and a high computational complexity. In this chapter, the use of a stochastic approach as Learning Automata (LA) to overcome these problems is explained. Few works have been reported using some stochastic methodology either to incorporate it into MPC or to generate a planning structure. An exception is reported in [15], through the use of a Probabilistic Neural-network as part of a MPC system applied to real time control of hybrid systems.

A learning automaton may be considered as a system which modifies its strategy on the basis of its experience in order to reach good optimization performances in spite of unpredictable changes in the environment where it operates. In other words, learning automata should, by collecting and processing current information regarding the environment, be capable of changing their structure and parameters as time evolves to achieve the desired goal or the optimal performance (in some sense).

The main motivation behind the use of LA as an optimization algorithm for planning systems is to use its capabilities of global optimization when dealing with multimodal surfaces. Using LA, the search for the optimum is done within a probability space rather than seeking within a parameter space as done by other optimization algorithms. Automata are referred to as an automaton, acting embedded into an unknown random environment. Such automaton improves its performance to obtain an optimal action.

On the other hand, an action is applied to a random environment and gives a fitness value to the selected action of the automata. The response of the environment is used by automata to select its next action. This procedure is continued to reach the optimal action.

LA has been used for solve different sorts of engineering problems. For instance, pattern recognition [16], adaptive control [17], signal processing [18], and power systems [19]. Recently, some effective algorithms have been proposed for multimodal complex function optimization based on the LA (see [18, 20–22]). Furthermore, it was shown experimentally that the performance of these optimization algorithms is comparable to, or better than the genetic algorithm (GA) in the reduction of the searching space and quick convergence as it is shown in [22]. This chapter employs the algorithm called *continuous action reinforcement learning automata* (CARLA) as LA approach.

The CARLA algorithm first introduced by Howell et al. [23], has demonstrated through empirical results its effectiveness on optimization tasks in a wide range of applications. In [18] the CARLA algorithm was used to simultaneously tune on-line the parameters of a PID-controller for an engine idle-speed system. On the other hand, in [24], is shown experimentally that CARLA's performance in the tuning the PID coefficients is superior to those obtained by Genetic algorithms (GA) and Particle Swarm Optimization (PSO).

This chapter also discusses how to emulate the functionality of planning in order to decide how to control a plant. The study focuses on typical plants considered in conventional control. The planning strategy is the MPC methodology, incorporating Learning Automata as the optimization algorithm. The use of a stochastic approach allows dealing appropriately with the multimodal problem of the error surface, as it accelerates the computation process and eliminates the controller complexity. The effectiveness of the methodology is tested over a nonlinear plant (surge tank plant) and compared with the results offered by the Levenberg–Marquardt algorithm [25]. The algorithm LM was chosen to be considered the most used in planning strategies applied to control, besides LM has the best balance between precision and speed.

The chapter is organized as follows: Sect. 11.2 presents a brief review of control planning strategies while Sect. 11.3 discusses on the foundations and theory of Learning Automata. In Sect. 11.4, the proposed approach is implemented using the surge tank plant as non-linear example. In Sect. 11.5, the experimental results are presented while Sect. 11.6 concludes the chapter.

11.2 Planning Strategy Design

The concept of planning is commonly understood following common sense grounds. For instance, when humans plan activities for the weekend or when a solution for a daily-life problem is discussed. The solution normally arises from a collection of actions to be followed aiming to achieve specific goals. Such kind of action-sorting can be named as an *action plan* and may fall into the following planning steps:

1. *Planning domain* Refers to the first representation of the problem to be solved (i.e. a model).
2. *Setting goals* Essential to planning to define the required behavior or overall aims.
3. *Sticking to the plan* Sometimes humans simply react to situations without considering the consequences of their actions. It is better to develop the plan, reaching the goals easily.
4. *Selecting a strategy* The selection of the plan commonly involves projections into the future by means of a model, which requires considering a variety of sequences of task and sub-goals to be executed. It is usually necessary an optimization algorithm, which obtains the best plan, starting from tests that are generated considering a partial model of the problem.
5. *Executing the plan* After the selection, it must be decided how to execute that plan.

This chapter focuses on plants that are typically considered for conventional control testing. On the other hand, planning systems are assumed as a piece of code which is able to emulate the way experts do planning to solve a given control problem. Note, however, that there exists an essential difference on how we thought of fuzzy and expert control, particularly because a planning system uses an explicit model of the plant. In this section, several issues surrounding the choice of such a model are discussed in detail.

11.2.1 Closed-Loop Planning Configuration

A generic planning system can be set on the architecture of a standard control system as shown in Fig. 11.1. Following the context of human planning and solving, the problem domain is the plant and its environment. There are measured outputs $y(k)$ at step k (variables of the problem domain that can be sensed in real time), control actions $u(k)$ (which can affect the problem domain), disturbances $d(k)$ (which represent random events that can affect the problem domain and hence the measured variable $y(k)$), and goal $r(k)$, which is called the reference input in conventional control terminology. The goal represents what is like to be achieved within the problem domain. There are closed-loop specifications to define the performance specifications and stability requirements. The types of plants which are considered in this section are as follows:

$$y(k+1) = f(x(k), u(k), d(k)) \tag{11.1}$$

Fig. 11.1 Closed-loop planning system

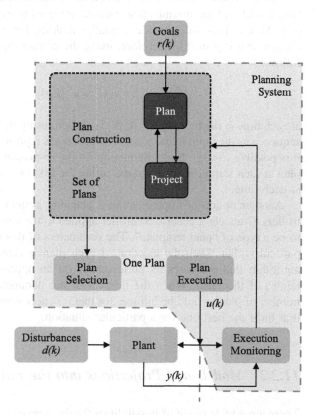

where $y(k)$ is the measured output and f is a generally unknown smooth function of the state $u(k)$ and measurable state $x(k)$.

$$x(k) = [y(k), y(k-1), \ldots, y(k-n), u(k-1), u(k-2), \ldots, u(k-m)]^T \quad (11.2)$$

Let

$$e(k) = r(k+1) - y(k+1) \quad (11.3)$$

Equation (11.3) is also known as the tracking error. Generally, the objective is to make the tracking error as small as possible all the time and asymptotically approaching zero, so that the output follows the reference input.

Considering a plan to be a sequence of possible control inputs and the ith-plan of length N at time k being structured as follows:

$$u^i[k, N] = u^i(k, 0), u^i(k, 1), \ldots, u^i(k, N-1) \quad (11.4)$$

The algorithm aims to develop a controller that is based on the planning strategy. One model and the optimization method are used to evaluate and score each plan (e.g. MPC). This will provide a quality ranking for each plan. The plan is thus chosen (call it plan $i*$). Therefore, using the control input at each time instant k as follows:

$$u(k) = u^{i*}(k, 0) \quad (11.5)$$

at each time k, the best plan $u^{i*}[k, N]$ is chosen, using the first input from the control sequence as the input to the plant. The process is repeated at each time step. Clearly, it is possible to use a lower frequency for the re-planning, using for instance a new plan at each sampling step, and executing the first two inputs from the optimal plan at each time.

Another option of implementing a planning system is the use of multiple controllers. Consider a generic controller applied to the current state an reference input to be a type of "plan template". The parameters of this generic controller specify a particular plan. Then the election of each plan is carried out by an optimization algorithm that evaluates its performance on an approximate model (with uncertainty) of the plant. When the interval of the parameters is continuous, then the number of plans would be infinite, for that it is necessary an optimization algorithm that finds the best one for a particular situation.

11.2.2 Models and Projections into the Future

There are a wide range of possibilities for the type of model that may be used. The type depends on the problem domain, the capabilities of the planner to store and use the model, and also the goals. For instance, a model used for planning could be continuous or discrete (e.g., a differential or difference equation), and it could be linear or nonlinear. It may be deterministic, or it may contain an explicit representation of the uncertainty of the problem, so the plans may also be chosen considering this factor [10].

Just like the design model used for controller construction, any given model cannot be a perfect representation of the plant and the environment, it will always be uncertainty in planning, and hence there will always be a bound on the amount of time required to sensibly simulate the model into the future. Such projecting into the future may become useless at some point should it goes too far because the predictions will become inaccurate at some point, providing poor information on how to select the best plan. The difficulty emerges from knowing how good your model may be and how far it may be projected into the future. In this chapter a deeper analysis on such problems will not be considered.

Here, we use a general nonlinear discrete time model

$$y_m(j+1) = f_m(x_m(j), u(j)) \tag{11.6}$$

with output $y_m(j+1)$, state $x_m(j)$, and input $u(j)$ for $j = 0, 1, 2,..., N-1$. Notice that this model can be quite general, if required. However in practical terms, a linear model is all that is available and this may be sufficient. Let $y_m^i(k,j)$ denote the jth value generated at time k using the ith plan $u^i[k, N]$, and $x_m(k,j)$. In order to predict the effects of plan i (as projected into the future) at each time k, it is required to calculate for $j = 0, 1, 2,..., N-1$, as follows

$$y_m^i(k,j+1) = f_m(x_m(k,j), u^i(k,j)) \tag{11.7}$$

At time k, simulating ahead of time, for $j = 0$, it begins with $x_m(k,0) = x(k)$. Then, it generates $y_m(k,j+1), j = 0, 1, 2,..., N-1$, using the model. It is required to appropriately shift values in x_m at each step and generate values of $u^i(k,j)$, $j = 1, 2,..., N-1$, for each i.

11.2.3 Optimization and Method for Plan Selection

Next, the set of plans (strategies) is "pruned" to one which is considered as the best one to apply at the current time. Hence, optimization is central to the planning. The specific type of optimization approach that is used for plan selection should be able to operate on multimodal surfaces, showing a light and fast computation. The previous requirements are usually difficult to solve by means of traditional optimization algorithms. This chapter therefore presents the use of LA as the optimization procedure.

Before the optimization procedure selection, it is necessary to define a specific criterion to decide on the best plan. In this chapter, a cost function of the type

$J(u^i[k, N])$ is used to quantify the quality of each candidate plan $u^i[k, n]$ by means of the model f_m. First, it is assumed that the reference input $r(k)$ is either known all the time or at least that at time k, while it is also known up to the time $k + N$. Therefore, the cost function is defined as follows:

$$J(u^i[k, N]) = \omega_1 \sum_{j=1}^{N} (r(k+j) - y_m^i(k, j))^2 + \omega_2 \sum_{j=1}^{N-1} (u^i(k, j))^2 \qquad (11.8)$$

being $\omega_1 > 0$ and $\omega_2 > 0$ scaling factors that are used to weight the importance of achieving the tracking error closely (first term) or minimizing the use of control energy (second term) to achieve that tracking error. In many cases ω_1 and ω_2 have the same value. To specify the control at time k, it is necessary simply to take the best plan, as it is measured by $J(u^i[k, N])$, and call it plan $u^{i^*}[k, N]$, while generating the control using $u(k) = u^{i^*}(k, 0)$ (i.e. the first control input in the sequence of inputs that was the best).

An important problem is therefore, to pick an optimization method that will converge to the optimal plan, and one that can cope with the complexity presented by the large number of candidate plans. First, focusing on the complexity aspect, notice that the inputs and states for the plant under consideration can, in general, take on a continuum number of values, even though in particular applications they may only take on a finite number of values. This is the case in analog control systems when considering actuator saturation. For digital control systems, one data acquisition system may be available, resulting into certain quantization and hence, theoretically speaking, there are a finite number of inputs, states, and outputs, for the model specified by f_m, since it is typically simulated on a digital computer. However, this number can be very large. In general, there exist an infinite (continuum) number of possible plans that must compute their cost, ranking all plans according to such cost and hence selecting the best one.

If non-linear and uncertain system characteristics dominate to the extent that a linear model is not sufficient for generating plans, then a nonlinear model can be used within the planner. Some type of nonlinear optimization method may therefore be used for the parameters that evaluate the infinite set of feasible plans. However, this may become a troublesome task since a non-linear model is used for plan generation, forcing the overall solution to consider non-linear optimization, with generally no analytical solution available.

There exists a wide variety of algorithms to tackle this problem such as steepest descent, Levenberg–Marquardt, etc. Such methods, however, do not guarantee convergence to an optimal plan or they may get stuck into local minima, generate divergent solutions or even provide no actual solution. Therefore the resulting plan after the non-linear optimization procedure cannot be assumed as optimal nor the closed-loop performance that it might generate, once it is included within one control loop. It is important to recall that for some practical industrial problems, engineers have managed to develop effective solutions via such a non-linear optimization approach, giving way to the main motivation beneath the use of LA as an optimization algorithm: it offers global optimization when dealing with multimodal surfaces. The search for the optimum is done within a probability space rather than seeking within a parameter space as it is done by other optimization algorithms. Automata are commonly portrait as an automaton, acting embedded into an unknown random environment and improving its performance to obtain an optimal action.

11.3 Learning Automata

The concept of learning automata was first introduced by the pioneering work of Tsetlin in the early 1960s [26]. Tsetlin was interested into the behaviour modelling of biological system. Subsequent research has considered the use of such learning paradigm for engineering systems.

LA operates by selecting actions via a stochastic process. Such actions operate within an environment while being assessed according to a measure of the system performance. Figure 11.2a shows the typical learning system architecture. The automaton selects an action ($\mathbf{X}$) probabilistically. Such actions are applied to the environment, and the performance evaluation function provides a reinforcement signal β. This is used to update the automaton's internal probability distribution whereby actions that achieve desirable performance are reinforced via an increased probability, while those not-performing actions are penalised or left unchanged depending on the particular learning rule which has been employed. Over time, the

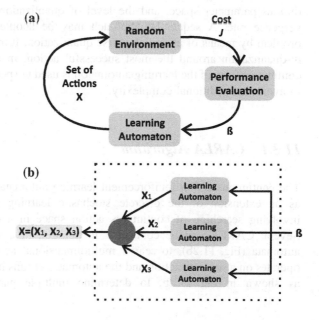

Fig. 11.2 **a** Reinforcement learning system and **b** interconnected automata

average performance of the system will improve while a given limit is reached. In terms of optimization problems, the action with the highest probability would correspond to the global minimum as demonstrated by rigorous proofs of convergence available in [27–29].

A wide variety of learning rules have been reported in the literature. One of the most widely used algorithms is the linear reward/inaction (L_{RI}) scheme, which has been shown to guaranteed convergence properties (see [27]). In response to action x_i, being selected at time step k, the probabilities are updated as follows:

$$p_i(k+1) = p_i(k) + \theta \cdot \beta(k) \cdot (1 - p_i(k))$$
$$p_j(k+1) = p_j(k) - \theta \cdot \beta(k) \cdot p_j(k), \quad \text{if } i \neq j \tag{11.9}$$

being θ a learning rate parameter and $0 < \theta < 1$ and $\beta \in [0, 1]$ the reinforcement signal; $\beta = 1$ indicates the maximum reward and $\beta = 0$ is a null reward. Eventually, the probability of successful actions will increase to become close to unity. In case that a single and foremost successful action prevails, the automaton is deemed to have converged.

With a large number of discrete actions, the probability of selecting any particular action becomes low and the convergence time can become excessive. To avoid this, Automata can be connected in a parallel setup as shown by Fig. 11.2b. Each automaton operates a smaller number of actions and the 'team' works together in a co-operative manner. This scheme can also be used where multiple actions are required.

Discrete stochastic learning automata can be used to determine global optimal states for control applications with multi-modal mean square error surfaces. However, the discrete nature of the automata requires the discretization of a continuous parameter space, and the level of quantization tends to reduce the convergence rate. A sequential approach may be adopted [18] to overcome such problem by means of an initial coarse quantization. It may be later refined using a re-quantization around the most successful action. In this chapter, an inherently continuous form of the learning automaton is used to speed the learning process and to avoid this additional complexity.

11.3.1   CARLA Algorithm

The continuous action reinforcement learning automata (CARLA) was developed as an extension of the discrete stochastic learning automata for applications involving searching of continuous action space in a random environment [18]. Several CARLA can be connected in parallel, in a similar manner to discrete automata (Fig. 11.2b), to search multidimensional action spaces. Each CARLA operates on a separate action and the automata set runs in a parallel implementation as shown in Fig. 11.2b, to determine multiple parameter values. The only interconnection between CARLAs is through the environment and via a shared performance evaluation function. The automaton's discrete probability distribution is replaced by a continuous probability density function which is used as the basis for action selection. It operates a reward/inaction learning rule similar to the discrete learning automata. Successful actions receive and increase on the probability of future selection via a Gaussian neighborhood function which increases the probability density in the vicinity of such successful action. The initial probability distribution can be chosen as being uniform over a desired range and over many iterations, this converges to a Gaussian distribution around the best action value.

If action x is defined over the range (x_{min}, x_{max}), the probability density function $f(x, n)$ at iteration n is updated according to the following rule:

$$f(x, n+1) = \begin{cases} \alpha \cdot [f(x, n) + \beta(n) \cdot H(x, r)] & \text{if } x \in (x_{min}, x_{min}) \\ 0 & \text{otherwise} \end{cases} \quad (11.10)$$

With α being chosen to re-normalize the distribution according to the following condition

$$\int_{x_{min}}^{x_{max}} f(x, n+1)dx = 1 \quad (11.11)$$

with $\beta(n)$ being again the reinforcement signal from the performance evaluation and $H(x, r)$ a symmetric Gaussian neighbourhood function centered on $r = x(n)$. It yields

$$H(x, r) = \lambda \cdot \exp\left(-\frac{(x-r)^2}{2\sigma^2}\right) \quad (11.12)$$

with λ and σ being parameters that determine the height and width of the neighborhood function. They are defined in terms of the range of actions as follows:

$$\sigma = g_w \cdot (x_{max} - x_{min}) \quad (11.13)$$

$$\lambda = \frac{g_h}{(x_{max} - x_{min})} \quad (11.14)$$

The value g_w controls the width of the Gaussian function that is added to the distribution in Eq. (11.10), while g_h the height. The speed and resolution of learning are thus controlled by free parameters g_w and g_h. Let action $x(n)$ be applied to the environment at iteration n, returning a cost or performance index $J(n)$. Current and previous costs are stored as a reference set $R(n)$. The median and minimum values J_{med} and J_{min} may thus be calculated, by means of $\beta(n)$ being defined as:

$$\beta(n) = \max\left\{0, \frac{J_{\text{med}} - J(n)}{J_{\text{med}} - J_{\text{min}}}\right\}$$ (11.15)

To avoid problems with infinite storage, and to allow the system to adapt to changing environments, only the last m values of the cost functions are stored in $R(n)$. Equation (11.15) limits $\beta(n)$ to values between 0 and 1 and only returns nonzero values for costs that are below the median value. It is easy to understand how $\beta(n)$ affects the learning process informally as follows: during the learning, the performance and the number of selecting actions can be wildly variable, generating extremely high computing costs. However, $\beta(n)$ is insensitive to these extremes and to the very high values of $J(n)$ resulting from a poor choice of actions. As learning continues, the automaton converges towards more worthy regions of the parameter space and these actions within such regions are chosen for evaluation increasingly often. While more of such responses are being received, J_{med} gets reduced. Decreasing J_{med} in the performance index effectively enables the automaton to refine its reference around the better responses previously received, and hence resulting in a better discrimination between the competing selected actions.

To define an action value $x(n)$ which has been associated to this probability density function, an uniformly distributed pseudo-random number $z(n)$ is generated within the range of $[0, 1]$. Simple interpolation is then employed to equate this value to the cumulative distribution function:

$$\int_{x_{\text{min}}}^{x(n)} f(x, n)dx = z(n)$$ (11.16)

In the CARLA optimization method, a probability density function is associated with each decision variable, and through modification of these probability density functions over sufficient number of iterations the optimal value of the decision variables will be determined. The modification process in each iteration is due to reinforcement signal $\beta(n)$ corresponding to a predefined cost function. For implementation purposes, the distribution is stored at discrete points with an equal inter-sample probability. Linear interpolation is used to determine values at intermediate positions (see full details in [18]).

11.4   Implementation

Our approach represents a planning system based on an approximate model of the plant that allows proving among different probable plans. The election of the best plan will be carried out by the LA considering its performance on the approximate model, projected forward in some instants. The election of each plan is made in each sampling instant k (as it was treated in Sect. 11.2.3). In this section, the planning strategy for a typical control plant the "surge tank" is designed. The discussion begins by introducing the control problem as it later moves to the designing and testing of the planning strategy.

11.4.1 Level Control in a Surge Tank

Consider the "surge tank", shown in Fig. 11.3 that can be modeled by

$$\frac{dh(t)}{dt} = -\frac{\bar{d} \cdot \sqrt{2gh(t)}}{A(h(t))} + \frac{\bar{c}}{A(h(t))} \cdot u(t) \qquad (11.17)$$

with $u(t)$ being the input flow (control input) which can be positive or negative (it can either pull liquid out of the tank or contribute to fill it in); $h(t)$ is the liquid level (the output of the plant); $A(h(t)) = |\bar{a} \cdot h(t) + \bar{b}|$ is the cross-sectional area of the tank with $\bar{a} > 0$ and $\bar{b} > 0$ (their nominal values are $\bar{a} = 0.01$ and $\bar{b} = 0.2$); $g = 9.8$; $\bar{c} \in [0.9, 1]$ is a "clogging factor" for a filter in the pump actuator, with $\bar{c} = 0.9$. There exists some clogging of the filter. However, in case $\bar{c} = 1$, the filter is clean so there is no clogging ($\bar{c} = 1$ will be taken as its nominal value); $\bar{d} > 0$ is a parameter related to the diameter of the output pipe (and its nominal value is $\bar{d} = 1$). It is assumed that all these plant parameters are fixed but unknown.

Let $r(t)$ be the desired level of the liquid in the tank (the reference input) and $e(t) = r(t) - h(t)$ be the tracking error. It is assumed the reference trajectory is known in advance and $h(0) = 1$. To convert to a discrete-time approach, the Euler approximation is used considering a sampling time of $T = 0.1$.

Fig. 11.3 The surge tank system

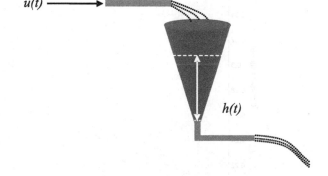

$u(t)$

$h(t)$

11.4.2   Planning System

For planning purposes, an uncertainty version of the nonlinear discrete-time model (the "truth model") is considered. The model can be considered as an approximation of the problem on which the different plans are proven. The candidate plans are generated using such a model while the evaluation follows the LA approach discussed in the last section. Taking the model of the last subsection as the truth model for the plant, the planning strategy model considers a different cross-sectional area, thus yielding:

$$A_m(h(t)) = \bar{a}_m(h(t))^2 + \bar{b}_m \qquad (11.18)$$

with $\bar{a}_m = 0.002$ and $\bar{b}_m = 0.2$. The same nonlinear equations in Eq. (11.16) are used assuming the values of $\bar{c}_m = 0.9$ and $\bar{d}_m = 0.8$. The Fig. 11.4 shows the cross-sectional area of the actual plant, and the one used in the model. It is evident in Fig. 11.4, the differences among the real plant data and the model data employed by the planning strategy.

In order to apply the planning methodology to the control of the given plant, a simple proportional integral (PI) controller is considered. In particular, if $e(k) = r(k) - h(\mathrm{k})$, it yields

$$u(k) = K_p \cdot e(k) + K_i \cdot \sum_{j=0}^{k} e(j) \qquad (11.19)$$

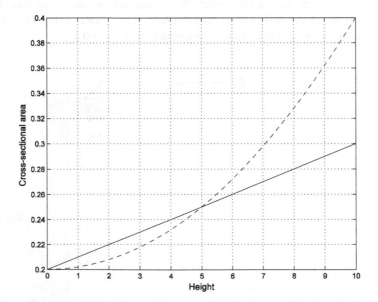

Fig. 11.4 Cross-sectional area $A(h)$ for the actual plant (*solid*) and the projected model (*dashed*)

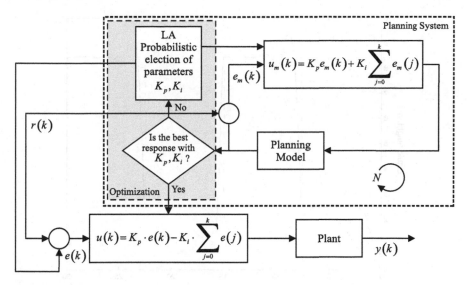

Fig. 11.5 Block representation of the system including the planning strategy

Each plan will be considered as a controller defined by its couple of coefficients and K_i, therefore the number of plans is infinite. The complete structure of the planning system contains the model which evaluates each plan and the optimization system which determines the best coefficients for the values—they must fulfill the acting indexes based on the error evaluation. Figure 11.5 shows a block representation of the system.

11.4.3 LA Optimization

As each plan is considered as a controller defined by its couple of coefficients K_p and K_i, then the problem will be to find the appropriate plan that represents the couple of coefficients with better performance, using the projection of the control action. Considering the optimization parameters, the intervals are chosen as $K_p \in [0, 0.2]$ and $K_i \in [0.15, 0.4]$. For instance, if the PI controller of Eq. (11.18) is calibrated using conventional techniques to control the plant, the values $K_p = 0.01$ and $K_i = 0.3$ are considered as optimal values. Figure 11.6 shows the controller performance. It is important to notice the fast response including overshoot which is not desirable.

K_p and K_i are not constant as they are calculated at each instant k by means of the optimization system (LA in this chapter). The LA optimization system chooses the parameters K_p and K_i according to a probability distribution. They are projected into the planning strategy. In case of an inconvenient result after the minimization

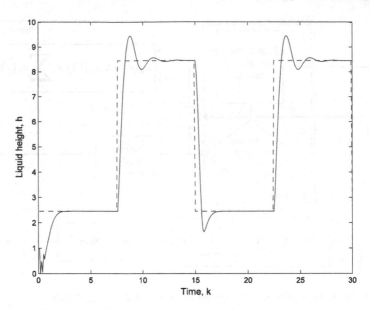

Fig. 11.6 Controller performance with $K_p = 0.01$ and $K_i = 0.3$

criterion is applied (performance index defined in Eq. (11.8), the probability distribution is modified. After several iterations, it must converge to a probability distribution around the optimal parameter value.

Equation (11.8) is used as minimization criterion, with $N = 20$ (for two projections into the future), $\omega_1 = 1$ and $\omega_2 = 1$. Also, the reference input is assumed to remain constant at each time. This is equivalent to assume that the evaluation of the best controller is based on the fact that the reference input is constant.

In the optimization process, two LA (one for each parameter) are used. They are coupled only through the environment (model). The set R is limited to 10 values while only 50 iterations of the CARLA learning are applied. The CARLA parameters are fixed at $g_w^{K_p} = 0.02$ and $g_h^{K_p} = 0.3$ for K_p, and $g_w^{K_i} = 0.02$ and $g_h^{K_i}$ for K_i.

Next, the overall CARLA algorithm for the optimization is described:

Step 1:	Set iteration $n = 0$
Step 2:	Define the action set $A(n) = \{K_p, K_i\}$ such that $K_p \in [0, 0.2]$ and $K_i \in [0.15, 0.4]$
Step 3:	Define $f(K_p, n)$ and $f(K_i, n)$ to be the probability density functions at iteration n
Step 4:	Initialize $f(K_p, n)$ and $f(K_i, n)$ to a uniform distribution between the defined limits
Step 5:	Repeat while $n \leq 50$
(a)	Use a pseudo-random number generator between 0 and 1, for each selected action $z_p(n)$ and $z_i(n)$
(b)	Select $K_p \in A(n)$ and $K_i \in A(n)$, considering that the area under the probability density function is $\int_0^{K_p(n)} f(K_p, n) = z_p(n)$ and $\int_{0.15}^{K_i(n)} f(K_i, n) = z_i(n)$.

(continued)

(continued)

(c)	Project the control over a 20 discrete time intervals
(d)	Evaluate the performance using Eq. (11.8)
(e)	Append to R and evaluate the minimum, J_{min}, and median, J_{med}, values of R
(f)	Evaluate $\beta(n)$ via Eq. (11.15)
(g)	Update the probability density functions $f(K_p, n)$ and $f(K_i, n)$ using Eq. (11.10)
(h)	Increment iteration number n

The learning system search in the two-dimensional parameter space of K_p and K_i, aiming in reducing the values for J in Eq. (11.8).

11.5 Results

In order to test the operation of the planning strategy, the complete system is simulated during 30 s assuming a pre-determined signal reference $r(k)$. Figure 11.6 shows the performance of the proposed approach applied to the plant considering different values for ω_1 and ω_2. It is easy to identify different responses depending upon the chosen value of ω_1 and ω_2.

Figure 11.7a results from setting $\omega_1 = 1$ and $\omega_2 = 1$. A slower rise-time can be seen in contrast to Fig. 11.6 which uses the PI controller. The system is able to tune the planning strategy by adjusting ω_1 and ω_2 so that there is a small overshoot and still a reasonable good rise-time. Figure 11.7b is obtained by setting $\omega_1 = 0.8$ and $\omega_2 = 0.8$ while Fig. 11.7c by setting $\omega_1 = 5$ and $\omega_2 = 1$.

Two CARLA automata were used; one for each parameter K_p and K_i respectively, each initialised using a uniform distribution. The election of each plan (values of K_p and K_i) is made in each sampling instant according to CARLA algorithm (see Sect. 11.4.3). The evolutions of two probability-density functions are shown in Fig. 11.8 considering two different sampling instants. The Fig. 11.8a shows the obtaining of the values of K_p and K_i for $k = 93$ while Fig. 11.8b for $k = 156$. It is straightforward to identify how the probabilities converge to a maximum probability through the iterations. The highest probability value of K_p and K_i will be used as parameter for the controller, just as defined by Eq. (11.19), for a specific time instant k.

To prove the performance of our approach we compare it with the gradient algorithm Levenberg–Marquardt. For the comparison we use the implementation given in [30]. The LM algorithm will minimize the Eq. (11.8), updating de parameters according to the following equation:

$$\theta(n+1) = \theta(n) - (\nabla \varepsilon(\theta(n) \cdot \nabla \varepsilon(\theta(n))^T + \Lambda(n))^{-1} \nabla \varepsilon(\theta(n)\varepsilon(\theta(n) \qquad (11.20)$$

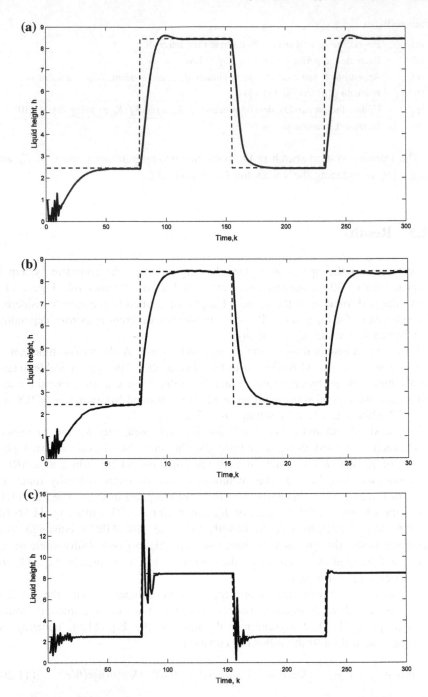

Fig. 11.7 Performance of the planning strategy applied to the plant, considering different values for ω_1 and ω_2. **a** Setting $\omega_1 = 1$ and $\omega_2 = 1$, **b** setting $\omega_1 = 0.8$ and $\omega_2 = 0.8$ and **c** setting $\omega_1 = 5$ and $\omega_2 = 1$

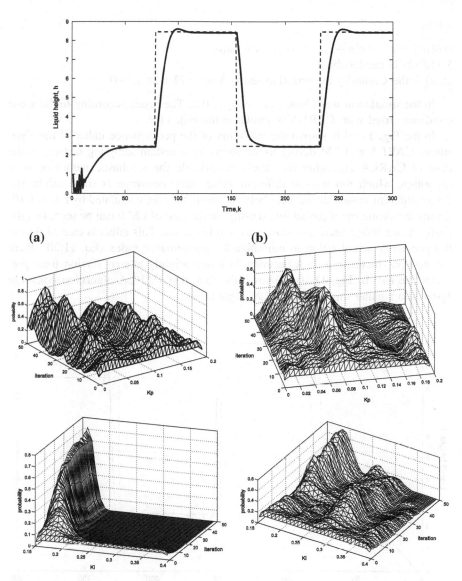

Fig. 11.8 Evolution of two probability-density functions obtained for two different operative points **a** $k = 93$ and **b** $k = 156$, considering $\omega_1 = 1$ and $\omega_2 = 1$

where

$\varepsilon(\theta(n)) = \omega_1 \cdot (r(n+j) - y_m(n)) + \omega_2 \cdot u(n)$
$\nabla \varepsilon(\theta(n))$ is the Jacobian
$\Lambda(n)$ is the Cholesky factorization term, $\Lambda(n) = \lambda \mathbf{I}$ with $\lambda > 0$

In the simulation was chosen $\omega_1 = \omega_2 = 0.8$. The result according to the same conditions used with CARLA is shown in the Fig. 11.9.

In the Fig. 11.10 is shown the evolution of the performance index of the algorithms CARLA and LM during its iterations for a certain sampling instant. In the case of CARLA algorithm its clearly remarkable the stochastic nature of such algorithm, which run tests at different values until converge at some minimum. These different values are not near between them, they are calculated by Eq. (11.10) during the evolution of probability density. In the case of LM it can be seen, that the performance index decreases slower as in other cases. This effect is caused during the parameter actualization to minimize the performance index (Eq. 11.20), there the new parameter $(n + 1)$ it becomes in a new slightly altered version from previous one (n). In Fig. 11.10 is shown the fact, that CARLA algorithm converge faster to a smaller minimum that LM algorithm

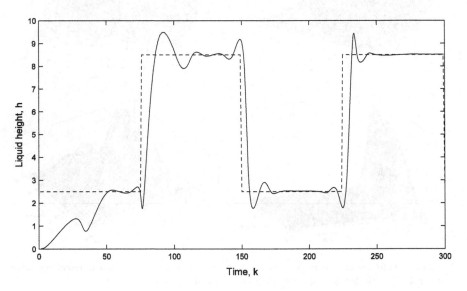

Fig. 11.9 Performance of the planning strategy using as optimization algorithm the Levenberg–Marquardt method

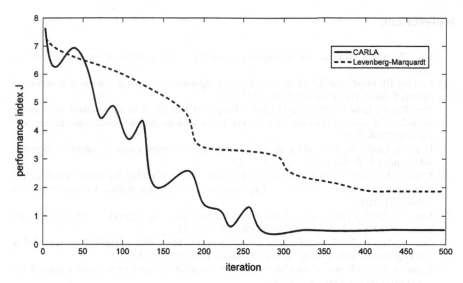

Fig. 11.10 Optimization results of the CARLA and Levenberg–Marquardt algorithms for a certain time instant k

11.6 Conclusions

This chapter has considered how to emulate the functionality of planning in order to decide how to control non-linear plants. The procedure adopts the MPC methodology as planning strategy while incorporating LA as the optimization algorithm. The effectiveness of the methodology was tested over a nonlinear plant and compared with the results offered by the Levenberg–Marquardt (LM) algorithm.

In this chapter, the LA is applied to select the optimal controller parameters. The MPC and the optimization algorithm run over an uncertain model of the plant. Following the results, the LA proves its abilities to choose the optimal parameters.

The approach is also suitable for real-time application. The CARLA algorithm has also been demonstrated to be applicable to adaptive systems such as in [18, 21, 22]. Although it requires learning in real-time, it can be effectively applied to nonlinear optimization problems which cannot be solved adequately by more conventional methods.

The proposed approach also allows increasing the optimization speed following a different approach from other optimization algorithms as LM. The search for the optimum points is performed within the probability space rather than inside the parameter space. Finally, the presented approach was compared to the LM algorithm, showing a faster and better convergence.

References

1. Liu GP (2001) Nonlinear identification and control: a neural network approach. Springer, Berlin
2. Fleming PJ, Purshouse RC (2002) Evolutionary algorithms in control systems engineering: a survey. Control Eng Pract 10(2002):1223–1241
3. Ying-Pin C, Low C, Shih-Yu H (2009) Integrated feasible direction method and genetic algorithm for optimal planning of harmonic filters with uncertainty conditions. Expert Syst Appl 36:3946–3955
4. Huang L (2009) Velocity planning for a mobile robot to track a moving target—a potential field approach. Robot Auton Syst 57:55–63
5. Chauvin J, Corde G, Petit N, Rouchon P (2008) Motion planning for experimental airpath control of a diesel homogeneous charge-compression ignition engine. Control Eng Pract 16(9):1081–1091
6. Son C (2006) Comparison of intelligent control planning algorithms for robot's part micro-assembly task. Eng Appl Artif Intell 19(1):41–52
7. Seow K, Sim K (2008) Collaborative assignment using belief-desire-intention agent modeling and negotiation with speedup strategies. Inf Sci 178(2):1110–1132
8. Camacho E, Bordons C (2008) Model predictive control (advanced textbooks in control and signal processing). Springer, Berlin
9. Garcia CE, Prett DM, Morari M (1989) Model predictive control: theory and practice—a survey. Automatica 25:335
10. Mayne DQ, Rawlings JB, Rao CV, Scokaert POM (2000) Constrained model predictive control: stability and optimality. Automatica 36(6):789
11. Camacho EF, Bordons C (2007) Nonlinear model predictive control: an introductory review. Assessment and future directions of nonlinear model predictive control. Springer, Berlin
12. Nagy ZK, Mahn B, Franke R, Allgower F (2007) Evaluation study of an efficient output feedback nonlinear model predictive control for temperature tracking in an industrial batch reactor. Control Eng Pract 15(7):839–850
13. Cueli R, Bordons C (2008) Iterative nonlinear model predictive control. Stability, robustness and applications. Control Eng Pract 16:1023–1034
14. Canale M, Fagiano L, Milanese M (2009) Set membership approximation theory for fast implementation of model predictive control laws. Automatica 45:45–54
15. Potočnik B, Mušič G, Škrjanc I, Zupančič B (2008) Model-based predictive control of hybrid systems: a probabilistic neural-network approach to real-time control. J Intell Robot Syst 51:45–63
16. Seyed-Hamid Z (2008) Learning automata based classifier. Pattern Recognit Lett 29:40–48
17. Zeng X, Zhou J, Vasseur C (2000) A strategy for controlling non-linear systems using a learning automaton. Automatica 36:1517–1524
18. Howell M, Gordon T (2001) Continuous action reinforcement learning automata and their application to adaptive digital filter design. Eng Appl Artif Intell 14:549–561
19. Wu QH (1995) Learning coordinated control of power systems using inter-connected learning automata. Int J Electr Power Energy Syst 17:91–99
20. Thathachar MAL, Sastry PS (2002) Varieties of learning automata: an overview. IEEE Trans Syst Man Cybern Part B Cybern 32:711–722
21. Zeng X, Liu Z (2005) A learning automaton based algorithm for optimization of continuous complex function. Inf Sci 174:165–175
22. Beygi H, Meybodi MR (2006) A new action-set learning automaton for function optimization. Int J Franklin Inst 343:27–47
23. Howell MN, Frost GP, Gordon TJ, Wu QH (1997) Continuous action reinforcement learning applied to vehicle suspension control. Mechatronics 7(3):263–276
24. Kashki M, Lofty Y, Abdel-Magid, Abido MA (2008) Advanced intelligent computing theories and applications. With aspects of artificial intelligence. In: Huang et al (ed)

A reinforcement learning automata optimization approach for optimum tuning of PID controller in AVR system. Springer, Berlin, pp 684–692

25. Kelley CT (2000) Iterative methods for optimization, SIAM frontiers in applied mathematics, no 18. ISBN 0-89871-433-8
26. Tsetlin ML (1973) Automaton theory and modeling of biological systems. Academic Press, New York
27. Narendra KS, Thathachar MAL (1989) Learning automata: an introduction. Prentice-Hall, London
28. Najim K, Poznyak AS (1994) Learning automata—theory and applications. Pergamon Press, Oxford
29. Thathachar MAL, Sastry S (2004) Techniques for online stochastic optimization. Springer, New York
30. Press W, Flannery B, Teukolsky, Vetterling W (1992) Numerical recipes in C: the art of scientific computing, 2 edn. Cambridge University Press, Cambridge

Chapter 12
Fuzzy-Based System for Corner Detection

Corner detection is an important task in computer vision problems due to the complexity of determinate the shape of different regions within an image. Real-life image data are always inexact due to inherent uncertainties that may arise from the imaging capture process such as defocusing, illumination changes, noise, etc. Therefore, the localization and detection of corners has become a difficult task under research, in order to accomplish the detection process under such imperfect situations. On the other hand, Fuzzy systems are well known for their efficient handling capacities when they face uncertainness and incompleteness. Fuzzy systems use modelling concepts in the same way as a human do. Under this circumstances, corners could be modelled by means of linguistic rules. This chapter presents a corner detection algorithm which employs fuzzy reasoning. The robustness of the presented algorithm is compared to well-known conventional corner detectors and its performance is also tested over a number of benchmark images to illustrate the efficiency of the algorithm under uncertainty.

12.1 Introduction

The human visual system has a highly developed capability for detecting many classes of patterns including visually significant arrangements of image elements. From the psychovisual aspect, points representing high curvature are one of the dominant classes of patterns that play an important role in almost all real life image analysis applications [1–3]. These points encode a significant amount of shape information. Corners are generally formed at the junction of different edge segments which may be the meeting (or crossing) of two edges. Cornerness of an edge segment depends solely on the curvature formed at the meeting point of two line segments. Corner detection is one of the fundamental tasks in computer vision and it can be regarded as a special type of feature segmentation. Extracted corners can

© Springer International Publishing AG 2017   239
M.-A. Díaz-Cortés et al., *Engineering Applications of Soft Computing*,
Intelligent Systems Reference Library 129, DOI 10.1007/978-3-319-57813-2_12

be used for measurement and/or recognition purposes. A large number of algorithms already exist in the literature. In particular, corner detection on gray level images can be classified into two main approaches. In the first approach, the gray level image is first converted into its binary version for extraction of boundaries using some thresholding technique. After a successful extraction of boundaries, the corners or the high curvature points are detected using directional codes or other polygonal approximation techniques [4]. In the second approach, the gray level image is taken directly as an input for corner detection. In this paper, the discussion is restricted to the second approach only. Most of the general-purpose detectors based on gray level, use either a topology-based or an auto-correlation-based approach. Topology based corner detectors, mainly use gradients and surface curvature to define the measure of cornerness. Points are marked as corners, if the value of cornerness exceeds some predefined threshold condition. Alternatively, a measure of curvature can be obtained using auto-correlation [5–9].

There exist several classical corner detection algorithms for estimating corner points. Such detectors are based on a local structure matrix which consists on the first partial derivatives of the intensity function. An clear example is the Harris feature point detector [10], which is based on a comparison: the measure of the corner strength—which is defined by the method and is based on a local structure matrix—is compared to an appropriately chosen concrete threshold. Another well-known corner detector is the smallest univalue segment assimilating nucleus (SUSAN) detector which is based on brightness comparison [11]. It does not depend on image derivatives. The SUSAN area will reach a minimum while the nucleus lies on a corner point. The effectiveness of the above-mentioned algorithms is acceptable. Recent studies such as [12] demonstrate that the Harris corner detector performs better for several circumstances in comparison to the SUSAN algorithm.

Data from natural images are always imprecise and noisy due to inherent uncertainties that may arise from the imaging process (such as defocusing, wide variations of illuminations, etc.). Thus, precise localization and detection of corners become difficult under such imperfect situations. On the other hand, fuzzy systems are well known for efficiently handling of impreciseness and incompleteness [13–15] due to imperfection of data. Therefore, it may result reasonable to model corner properties using a fuzzy rule-based system as they have been successfully applied to image processing in a wide variety of applications [16–18]. This chapter seeks to contribute to enhance the application of fuzzy logic to image processing, just as it has been proposed in [19]. The method adopts a template-based rule-driven procedure and has been specifically developed to deal with topics related to image processing purposes. The presented method is able to address many different processing tasks [20–22] and to produce better results than classical methods when applied to some critical issues such as noise [20, 23, 24].

Only few fuzzy approaches have specifically addressed the problem of corner detection for general purposes. Banerjee and Kundu have proposed in [25] an algorithm to extract significant gray level corner points. The measure of cornerness in each point is computed by means of the fuzzy edge strength and the gradient direction. Different corner fuzzy-sets are obtained by considering different threshold values from the fuzzy edge map. However, the algorithm's main drawback is that it uses several feature detectors which operate at different stages, yielding a high computational load.

On the other hand, Várkonyi-Kóczy have proposed in [26], a fuzzy corner detector that employs a local structure matrix. It builds a continuous transient between the localized and not localized corner points. The algorithm uses a fuzzy pre-filter that improve the quality of the image under process. Despite both fuzzy approaches show a good performance, they demand an expensive computing load in comparison to other classical algorithms such as the Harris method or SUSAN.

This chapter presents a new robust algorithm to extract significant gray level corner points. The method is derived from a fuzzy-rule approach which aims to detect corners even under complex conditions. In the presented approach, the measure of "cornerness" for each pixel in the image is computed by fuzzy rules (represented as templates) which are applied to a set of pixels belonging to a rectangular window. As the algorithm scans each pixel of the image at a time, a new pixel of the resulting image is generated by fuzzy reasoning. Hence, the possible uncertainty contained in the window-neighborhood is handled by using an appropriate rule base (template set). Experimental evidence shows the effectiveness of such method for detecting corners under several conditions. A comparison between one state-of-the-art fuzzy-based method [25] and the Harris algorithm [10] demonstrates the performance of the Fuzzy-based method.

This chapter is organized as follows: Sect. 12.2 briefly describes the mathematical approach and the fuzzy model used in this work. Section 12.3 describes the experimental results while Sect. 12.4 describes the performance comparison regarding to other methods. On the other hand, Sect. 12.5 offers some conclusions about the development and performance of this technique.

12.2 Fuzzy Rule-Based System

12.2.1 Fuzzy System

Most of the approaches for corner detection are easy to implement and demand a low computational load. However, their effective operation largely relies on the fact that noisiness must be limited. In this section, a more robust technique is proposed. The new procedure is set to deliver a better performance for noisy environments. The fuzzy system is simple to implement and still fast in computation if it is compared to some existing fuzzy methods [25, 26]. Also, it can be easily extended to detect other features. In the proposed approach, the fuzzy rules are applied to a set of pixels belonging to a rectangular $N \times N$ window (usually 3×3 pixels), where the gray-level differences between the center pixel $p_{m,n}$ and its surrounding pixels are computed and stored within matrix E as follows:

$$E = \begin{bmatrix} p_{m,n} - p_{m-1,n-1} & p_{m,n} - p_{m-1,n} & p_{m,n} - p_{m-1,n+1} \\ p_{m,n} - p_{m,n-1} & 0 & p_{m,n} - p_{m,n+1} \\ p_{m,n} - p_{m+1,n-1} & p_{m,n} - p_{m+1,n} & p_{m,n} - p_{m+1,n+1} \end{bmatrix}, \qquad (12.1)$$

where m and n represent the coordinates of the central pixel. If the neighborhood is a homogenous region, then E contains values near zero. In the case of corners, the matrix E possesses a specific configuration depending on the corner type. These divide E in two connected regions, one with positive (pixel type A) and another with negative (pixel type B) difference values (see Fig. 12.1). The reasoning structure uses two different types of rules: the **THEN-rules** and the **ELSE-rules** (don't care conditions) respectively. Each THEN-rule includes a determined pixel configuration as antecedent and only one pixel as consequent. Antecedents are related to a corner existence test and the consequent to its presence or absence. The rule-base gathers many fuzzy rules (THEN-rules) and only one ELSE-rule (i.e. do-not-care rule). Therefore, only relevant rules (i.e. configurations) are formulated as THEN-rules while other not important configurations may be handled as a group of ELSE-rules.

The set of THEN-rules lies on the very core of the algorithm. The rules must deliver successful structure detection, i.e. corners in this case, while still cancelling other inconsistencies such as noise. Such tradeoff may be solved by using a reduced set of rules (configurations) which in turn represent the minimum number in order to coherently detect the structure as it is required by a given application. Such procedure allows dealing with noisy pixels or imprecision.

The proposed corner detector considers twelve THEN-rules that represent the same number of possible corner configurations and only one ELSE-rule as it is graphically explained by Fig. 12.2. It may be also possible to consider some other corner configurations. However it may reduce the algorithm's ability to deal with noise or uncertainty [19, 20, 24].

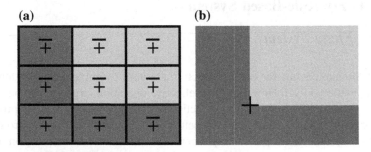

(a)   **(b)**

Fig. 12.1 Region shaping with respect to *gray* level differences: **a** the resulting template and **b** the real corner that originates the template

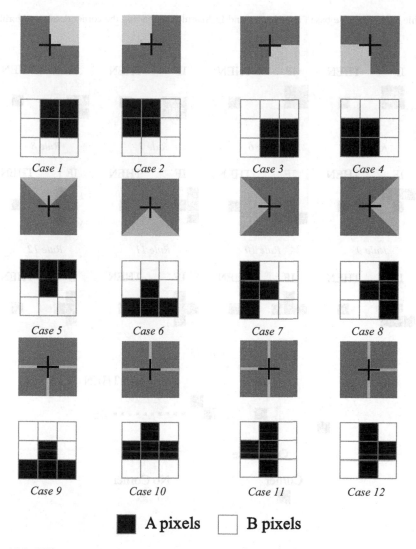

Fig. 12.2 Different corner cases to be considered for building the fuzzy rules. The image region containing the corner is shown in the *upper* section while the resulting 3 × 3 template is shown below each case

Despite using a reduced rule base, the performance in the detection process can be considered acceptable when it is compared to other algorithms solving the same task. The rule base (THEN-rules and ELSE-rule) supporting the detector algorithm is shown in Table 12.1.

Each rule has the following form:

Table 12.1 The rule base (THEN-rules and ELSE-rule) supporting the corner detector algorithm

Rule 1	Rule 2	Rule 3	Rule 4
IF THEN	IF THEN	IF THEN	IF THEN

Rule 5	Rule 6	Rule 7	Rule 8
IF THEN	IF THEN	IF THEN	IF THEN

Rule 9	Rule 10	Rule 11	Rule 12
IF THEN	IF THEN	IF THEN	IF THEN

Rule 13
IF   THEN

Otherwise
Corner   No Corner

If the corner structure in E possesses positive elements

and the opposite region possesses negative elements,

then the pixel represent a corner,

else the pixel does not represent a corner

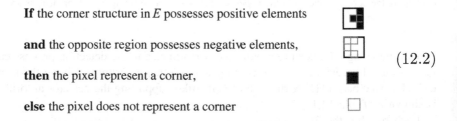

$$(12.2)$$

The principle can be explained as follows: If one region of the neighborhood, according to any of the twelve cases, contains positive/negative differences with respect to the center pixel, and if any other region contains the opposite (negative/positive) differences with respect to the center pixel, then the center pixel is a corner (see Fig. 12.2). The procedure can be considered as the evaluation of each one of the 12 different THEN-rules (configurations), yielding two auxiliary matrices E^p and E^n as follows:

$$E^p(i,j) = \begin{cases} 1 & \text{if } E(i,j) \geq 0 \\ 0 & \text{else} \end{cases}, \tag{12.3}$$

$$E^n(i,j) = \begin{cases} 1 & \text{if } E(i,j) < 0 \\ 0 & \text{else} \end{cases}, \tag{12.4}$$

where i, j represents de row and column of the matrix $E(i, j \in (1,2,3))$, Eq. 12.1.

For the case that all elements of E^p/E^n are ones (meaning all elements of $E(i,j)$ are positives or negatives), it is possible to construct regions A and B within the window-neighborhood according to the existing relative differences. Thus, the values of E^p and E^n can be recalculated as follows:

$$E^p(i,j) = \begin{cases} 1 & \text{if } E(i,j) \leq t_h \\ 0 & \text{else} \end{cases},$$
$$E^n(i,j) = \begin{cases} 1 & \text{if } E(i,j) > t_h \\ 0 & \text{else} \end{cases}. \tag{12.5}$$

For all the elements of E^p being ones, and

$$E^p(i,j) = \begin{cases} 1 & \text{if } E(i,j) \geq -t_h \\ 0 & \text{else} \end{cases},$$
$$E^n(i,j) = \begin{cases} 1 & \text{if } E(i,j) < -t_h \\ 0 & \text{else} \end{cases}. \tag{12.6}$$

For all the elements of E^n being ones, t_h is a threshold that controls the sensitivity of the considered differences. Typical values for t_h normally fall into the interval (5, 35). The lowest value of 5 would yield a higher detector's sensitivity which may detect a great number of corners corresponding to noisy intensity changes which are commonly found in images.

On the other hand, a maximum value of 35 would detect corners matching to a significant difference between several objects in the structure, i.e. object whose pixels may be considered as being connected. Although the selection of the best value for t_h clearly depends on the particular application, a good compromise can be obtained by taking a value on approximately half the overall interval, i.e. $t_h = 20$.

The membership values $\mu_c(m,n)$ (where $c = 1, 2, \ldots, 12$) are computed depending on the corner types (see Fig. 12.2). According to [27], such values represent the antecedents of each employed THEN-rule. They can be calculated as follows:

$$\mu_c(m,n) = \frac{1}{20}\max\left[\left(\sum_{ij\in A} E^p(i,j)\right)\cdot\left(\sum_{ij\in B} E^n(i,j)\right), \left(\sum_{ij\in B} E^p(i,j)\right)\cdot\left(\sum_{ij\in A} E^n(i,j)\right)\right].$$
(12.7)

Expression (12.7) considers a normalization factor equal to 20 which represents the maximum possible value, i.e. the highest product of the multiplication among the pixels between E^p and E^n. Hence, the membership value $\mu_c(m,n)$ falls between 0 and 1. Equation (12.7) can be considered as the numerical implementation of the generic rule previously defined by Eq. 12.2. If Rule 1 (case 1) is considered as an example, the expressions corresponding to expression (12.7) would thus be:

$$\sum_{ij\in A} E^p(i,j) = E^p(1,2) + E^p(1,3) + E^p(2,2) + E^p(2,3)$$

$$\sum_{ij\in B} E^n(i,j) = E^n(1,1) + E^n(2,1) + E^n(3,1) + E^n(3,2) + E^n(3,3)$$

$$\sum_{ij\in B} E^p(i,j) = E^p(1,1) + E^p(2,1) + E^p(3,1) + E^p(3,2) + E^p(3,3)$$

$$\sum_{ij\in A} E^n(i,j) = E^n(1,2) + E^n(1,3) + E^n(2,2) + E^n(2,3).$$
(12.8)

Analogously to (12.8), membership values $\mu_2(i,j), \ldots, \mu_{12}(i,j)$ for other rules (cases) can be calculated. Finally, the 12 fuzzy rules can be added into a single fuzzy value using the **max** (maximum) operator. The final fuzzy value represents the linguistic meaning of cornerness yielding:

$$\mu_{cornerness}(m,n) = \max(\mu_1(m,n), \mu_2(m,n), \ldots, \mu_{12}(m,n)) \qquad (12.9)$$

The pixels whose value $\mu_{cornerness}(m,n)$ are near to one, belong to a feature similar to a corner, while values near to zero would represent any other feature.

12.2.2   Robustness

This kind of corner detection clearly differs from other classical procedures in several ways. Conventional corner detectors look usually for the explicit corner location by means of detecting the zero-crossing of derivatives in different

Fig. 12.3 The effect of the impulsive noise in matrices E^+ and E^-. Matrices E^+ and E^- would contain only ones or zeros depending on the gray-level difference

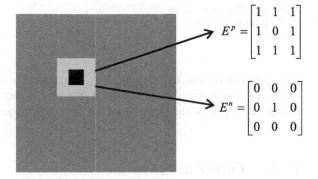

$$E^p = \begin{bmatrix} 1 & 1 & 1 \\ 1 & 0 & 1 \\ 1 & 1 & 1 \end{bmatrix}$$

$$E^n = \begin{bmatrix} 0 & 0 & 0 \\ 0 & 1 & 0 \\ 0 & 0 & 0 \end{bmatrix}$$

directions. On the contrary the presented approach detects the entire area where the corner could lie.

In particular, gradient-based methods are normally highly sensitive to the noise in real images and being mainly affected by the impulsive noise. Also, most of the corner detection algorithms incorporate several pre-filters [1, 10, 11, 28], which allow attenuation but do not eliminate impulsive noise.

On the other hand, fuzzy detectors allow corner marking despite noisy environments either by implementing fuzzy pre-filtering that eliminates uncertainty on the image or by incorporating fuzzy sets for modeling imprecision [25]. The method presented in this chapter considers vagueness due to noise and grayness ambiguity to be handled by the fuzzy rules introduced in expression (12.2). If the image of Fig. 12.3 is considered, with a pixel holding a gray value different from its neighbors is located within a homogeneous region. This situation can be considered as impulsive noise.

Under these circumstances, matrices E^p and E^p would contain only ones or zeros depending on the gray-level difference. Therefore, the values used to calculate the membership functions in expression (12.7), for any of the twelve cases, would yield

$$\left(\sum_{ij \in Rp} E^p(i,j) \right) \cdot \left(\sum_{ij \in Rn} E^p(i,j) \right) \approx 0, \qquad (12.10)$$

$$\left(\sum_{ij \in Rn} E^p(i,j) \right) \cdot \left(\sum_{ij \in Rp} E^n(i,j) \right) \approx 0. \qquad (12.11)$$

Now, considering the values from expressions (12.10) to (12.11) and a noisy pixel, the resulting value of its cornerness can be calculated by expression (12.9) as $\mu_{cornerness}(i,j) \approx 0$. The impulsive noise is thus classified by the fuzzy system as a

Table 12.2 Parameter setup for the proposed corner detector	t_h	t_c	H
	20	0.7	10

homogeneous region. In the same way, the central pixel would not be marked as corner for cases not considered in Table 12.2 which normally represent noisy configurations. It is mainly because the inference system works with ELSE-rules.

12.2.3   Corner Selection

In order to detect corners, it would be enough to choose an appropriate threshold t_c. If $\mu_{cornerness}(m,n) \geq t_c$, then the pixel $p_{m,n}$ can be assumed as such. Under these assumptions, the value t_c must be selected as close to 1 as it is likely to assure that pixel $p_{m,n}$ may be a corner. However, a more convenient approach is to choose a small threshold value t_c whose value allows detecting a wider number of corners despite a higher uncertainty. The corner selection process can therefore be explained as follows: For each pixel, if $\mu_{cornerness}(m,n) \geq t_c$, a neighborhood of $H \times H$ dimension is established around it (commonly $H > N$). The pixel $p_{m,n}$ is thus selected as a corner if its value $\mu_{cornerness}(m,n)$ is maximum within the neighborhood $H \times H$, otherwise it does not represent a corner point.

Figure 12.4 shows a selection example, where $\mu_{cornerness}(m,n)$ represents the cornerness of the pixel currently under evaluation, by assuming $\mu_{cornerness}(m,n) \geq t_c$. Inside the window $H \times H$ that has been established around it, there exist other two pixels $p_{i,j}$ and $p_{i',j'}$, whose values $\mu_{cornerness}(i,j)$ and $\mu_{cornerness}(i',j')$ are lower than $\mu_{cornerness}(m,n)$. Therefore, a point $p_{m,n}$ can thus be considered as a corner within the image.

12.3   Experimental Results

Different sorts of images have been tested in order to analyze the performance of the method for corner detection. Such benchmark set includes image alterations such as blurring, illumination change, impulsive noise etc. Table 12.2 presents the parameters of the presented algorithm used in this chapter. Once they have been determined experimentally, they are kept for all the test images through all experiments.

Fig. 12.4 Neighborhood
method for corner selection

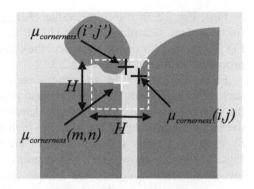

Fig. 12.5 a Detected corners
using the proposed approach,
and **b** values of
$\mu_{cornerness}(m,n)$

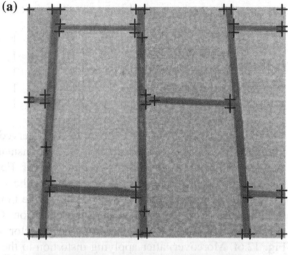

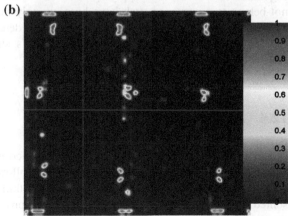

First, Fig. 12.5b shows the value of $\mu_{corneress}(m, n)$, as it is computed by the fuzzy system according to Eq. (12.9) to detect corners in a real image. In Fig. 12.5a, the blue crosses represent the corners obtained using the corner selection procedure explained in Sect. 12.2.3.

Figure 12.6 shows the algorithm's performance on different image conditions such as the case with variable illumination and blurring. Figures 12.6a, b present the performance of the fuzzy corner detector as it is applied to over-exposed and over-illuminated images. The effect of high illumination on the images was made by applying a linear transformation of the form $I(i, j) + 80$. On the other hand, Figs. 12.6c, d show the effectiveness of the presented detector using low-illuminated or sub-exposed images. Such effect was made by another linear transformation: $I(i, j) - 40$. The images in Fig. 12.6e, f illustrate the sensitivity of the fuzzy detector to blurring. Such steamed up effect was made by applying a low-pass filter to the original images, with a 5×5 kernel as follows:

$$h(i, j) = \frac{1}{25} \begin{bmatrix} 1 & 1 & 1 & 1 & 1 \\ 1 & 1 & 1 & 1 & 1 \\ 1 & 1 & 1 & 1 & 1 \\ 1 & 1 & 1 & 1 & 1 \\ 1 & 1 & 1 & 1 & 1 \end{bmatrix}. \tag{12.12}$$

From results shown in Fig. 12.6, it can be observed as the fuzzy detector exhibits immunity to changes in illumination, see for instance Fig. 12.6a–d. However, it also shows sensitivity to blurring in Fig. 12.6e, f. For the case of blurring images, the detector is able to find all the corners over the simulated image in 12.6e.

The latter figure exhibits low distortion in the homogeneous gray levels within the image as a consequence of the filter operation. On the other hand, some sensitivity may be lost while applying the detector to the real image shown in Fig. 12.6f. Moreover, after applying distortion to the image, several points that do not belong to a corner as such have been wrongly marked as corners.

Despite all previous comments, the fuzzy detector was able to detect in Fig. 12.6f the corners which delimit the object's shape. This is not a common feature of other corner detectors [9–12].

12.4 Performance Comparison

A variety of quantitative evaluation methods for corner detection algorithms have been proposed in the literature [12, 29, 30]. Following the criteria in [30], the performance analysis considers the Harris algorithm [10], the fuzzy method presented by Banerjee and Kundu [25] and the approach presented in this chapter.

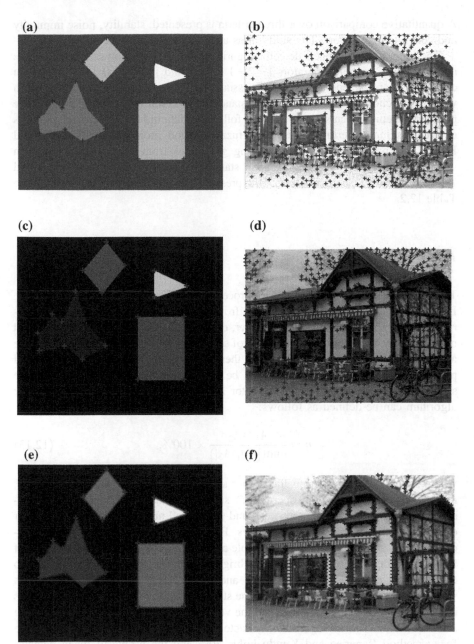

Fig. 12.6 Performance of the fuzzy corner detector over different conditions on the image: **a**, **b** over-exposition or high illumination **c**, **d** sub-exposition or low illumination and **e**, **f** blurring

A quantitative comparison over three criteria is presented: stability, noise immunity and computational effort. The study aims analyze the performance objectively.

The parameters for each detector algorithm are set as follows: For the Harris algorithm, the gradient operators $[- 2 -1\ 0\ 1\ 2]$ and $[- 2 -1\ 0\ 1\ 2]^T$ are set in directions u and v separately. The Gaussian smoothing filter employs a Gaussian window function of size 7×7 and a standard deviation of 2 with $k = 0.06$. The parametric setup appears as the best set following data in [12] and considering lots of hand tuning experiments. For the fuzzy method proposed by Banerjee and Kundu, the parameter are set following guidelines from [25], with a Gaussian window function of size 3×3 and a standard deviation of 2, $\mu_d(P) \geq 0.9$ and $T_h = 0.2$. Finally, the parameters of the presented approach are set according to the Table 12.2.

12.4.1 Stability Criterion

Two frames in an image sequence are processed by the algorithm to detect corners. If the corner's positions are unchanged from one frame to the next one, the algorithm can be regarded as stable. However, the gray-level value of each pixel would normally vary in actual images because of several factors affecting the image. If the algorithm is applied to a given image, then it cannot be assured the number and position of all detected corners would be exactly the same. Therefore, absolute stability is almost non-existent. A factor η to measure the stability of a corner algorithm can be defined as follows:

$$\eta = \frac{A_1 \cap A_2}{\min(|A_1|, |A_2|)} \times 100\%, \tag{12.13}$$

where A_1 and A_2 representing the corner sets for the first and the second frame, respectively (the intersection operator $\cap$ stands for common corners); $|A_i|$ represents the number of elements in A_i set and the overall numerator holds the number of corresponding corners in two frames. From expression (12.8), it can be concluded that a greater η yields a more stable corner detection algorithm. Fifty pairs of images holding different contrast and brightness levels are gathered in order to compare the presented fuzzy detector and other classic methods. Figure 12.7a shows the comparison with respect to the stability factor, where the horizontal axis represents the image pair number and the vertical axis represents the value of such stability factor. The average stability factor of Harris detector is 75%, while the fuzzy method Banerjee and Kundu holds 70% and the proposed fuzzy detector shows 83%.

12.4.2 Noise Immunity

Noise immunity is measured by factor ρ which it can be defined as follows:

$$\rho = \frac{B_1 \cap B_2}{\max(|B_1|, |B_2|)} \times 100\%, \qquad (12.14)$$

where B_1 is the corner set of the original image and B_2 is the corner set of the image with added noise. In this case, the maximum operator seeks to consider that false corners have been added as a result of additive noise. As ρ increases, it can be assumed that the algorithm's ability to avoid noisy corners is stronger.

One experiment is focus on comparing such noise immunity among methods. Fifty images with 10% of added impulsive noise are considered. Figure 12.7b

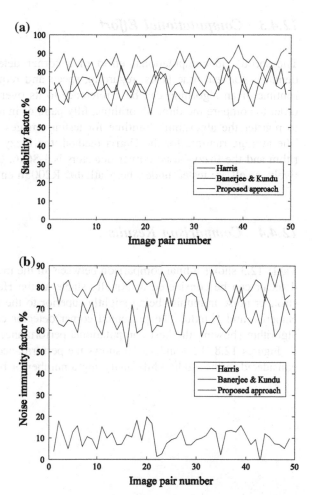

Fig. 12.7 Performance comparison among corner detectors. **a** Stability factor and **b** noisy immunity factor

Table 12.3 Performance comparison among the three corner detectors considered by the study

Corner detector	Stability ± standard deviation (%)	Noise ± standard deviation (%)	Time ± standard deviation (s)
Harris	75 ± 5.5	9 ± 4.4	1.8686 ± 0.3
Fuzzy Banerjee and Kundu	70 ± 7.8	65 ± 7.1	6.2125 ± 0.21
The presented fuzzy-based detector	83 ± 4.1	80 ± 4.6	0.878 ± 0.11

shows the noise immunity factor, with the Harris detector showing 9%, the fuzzy method Banerjee and Kundu holding 65% and the proposed fuzzy detector showing 80%.

12.4.3 Computational Effort

The speed and computational effort of a corner detector algorithm must meet demands for real-time tasks, regarding speed and required processing time. The runtime of an algorithm can be a reference to its overall computational effort. In order to compare the three algorithms, fifty pairs of images are considered in order to register the algorithm's runtime for testing images holding 320×240 pixels. The average runtime for the Harris method, the fuzzy Banerjee and Kundu algorithm and the fuzzy-based corner detectors is 1.8686, 6.2125 and 0.878 s, respectively, as all are tested under the MatLab© R2008b environment.

12.4.4 Comparison Results

Table 12.3 shows a final comparison between all the methods. The presented fuzzy detector can be considered as equally stable as the Harris method. It also shows stronger noise immunity being slightly superior to the fuzzy detector proposed by Banerjee and Kundu. The presented corner detector can also be regarded as the algorithm showing the best computational performance.

Figures 12.8, 12.9 and 12.10 shows the performance of the detector algorithms considered in the study while analyzing a number of benchmark images.

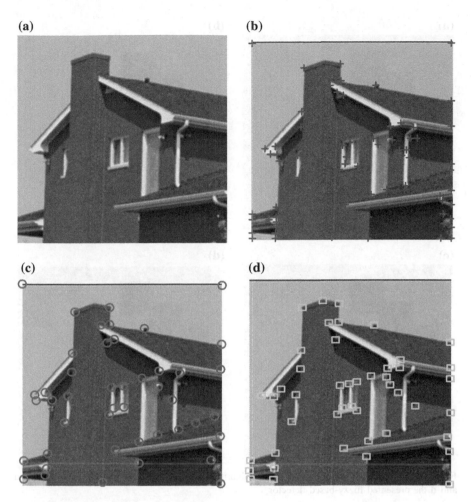

Fig. 12.8 House: **a** original image, **b** the Harris algorithm, **c** the fuzzy Banerjee and Kundu method and **d** the fuzzy-based presented detector

12.5 Conclusions

This chapter has presented a corner detection algorithm which models the structure of a potential corner in images based on a fuzzy rule set. The method is able to tolerate implicit imprecision and impulsive noise.

Experimental evidence suggests that the fuzzy-based presented algorithm produces better results than other common methods such as the Harris detector [10] and the fuzzy approach proposed by Banerjee and Kundu [25]. The presented algorithm is able to successfully identify corners on images holding different uncertainty conditions. However, it is also sensitive to blurring in particular when a

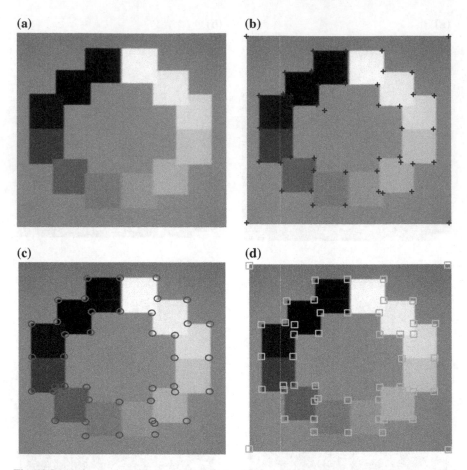

Fig. 12.9 Circle: **a** original image, **b** Harris algorithm, **c** the fuzzy Banerjee and Kundu method and **d** the presented fuzzy-based detector

steaming up effect is produced by considering neighborhood window wider than the one previously considered for building the fuzzy model of corners (templates). Such fact shall not be considered as inconvenient because the fuzzy-based algorithm is still capable of identifying corners over similar blurring levels than those of conventional algorithms.

The presented detector is stable and has shown robustness to impulsive noise which in turn represents its major advantage over the Harris method considering that impulsive noise is commonly found in real-time images. Although the algorithm exhibits a tolerance to imprecision that matches the performance of the Banerjee and Kundu fuzzy method, the presented approach requires a lighter computational cost for analyzing benchmark images.

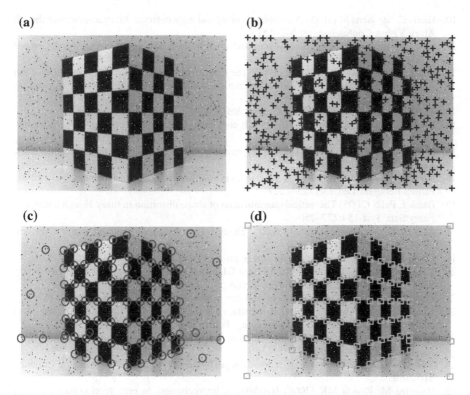

Fig. 12.10 Noisy chessboard image: **a** original image and the output after applying, **b** the Harris algorithm, **c** the fuzzy Banerjee and Kundu method and **d** the presented fuzzy-based detector

References

1. Lowe DG (1985) Perceptual organization and visual recognition. Kluwer Academic Publishers, USA
2. Loupias E, Sebe E (2000) Wavelet-based salient points: applications to image retrieval using color and texture features, in advances in visual information systems. In: Proceedings of the 4th international conference, VISUAL 2000, pp 223–232
3. Fischler M, Wolf HC (1994) Locating perceptually salient points on planar curves. IEEE Trans Pattern Anal Mach Intell 16(2):113–129
4. Freeman H, Davis LS (1977) A corner-finding algorithm for chain-coded curves. IEEE Trans Comput C-26:297–303
5. Kitchen L, Rosenfeld A (1982) Gray-level corner detection. Pattern Recogn Lett 1:95–102
6. Zheng Z, Wang H, Teoh E (1999) Analysis of gray level corner detection. Pattern Recogn Lett 20(2):149–162
7. Rattarangsi A, Chin RT (1992) Scale-based detection of corners of planar curves. IEEE Trans Pattern Anal Mach Intell 14(4):430–449
8. Teh C, Chin RT (1989) On the detection of dominant points on digital curves. IEEE Trans Pattern Anal Mach Intell 11(8):859–872
9. Rosenfeld A, Johnston E (1973) Angle detection on digital curves. IEEE Trans Comput C-22:858–875

10. Harris C, Stephens M (1988) A combined corner and edge detector. In: Proceedings of the 4th Alvey Vision Conference, pp 147–151
11. Smith S, Brady M (1997) A new approach to low level image processing. Int J Comput Vision 23(1):45–78
12. Zou L, Chen J, Zhang J, Dou L (2008) The Comparison of Two Typical Corner Detection Algorithms. Second international symposium on intelligent information technology application (ISBN 978-0-7695-3497)
13. Zadeh LA (1965) Fuzzy sets. Inf Control 8:338–353
14. Pal SK, Ghosh A, Kundu MK (2000) Soft computing for image processing. Physica-Verlag, Philadelphia, pp 44–78 (Chapter 1)
15. Yua D, Hu Q, Wua C (2007) Uncertainty measures for fuzzy relations and their applications. Appl Soft Comput 7(3):1135–1143
16. Karmakar G, Dooley L (2002) A generic fuzzy rule based in image segmentation algorithm. Pattern Recognit Lett 23:1215–1227
17. Basak J, Pal S (2005) Theoretical quantification of shape distortion in fuzzy Hough transform. Fuzzy Sets Syst 154:227–250
18. Jacquey F, Comby F, Strauss O (2008) Fuzzy Edge detection for omnidirectional images. Fuzzy Sets Syst 159:1991–2010
19. Russo F (1999) FIRE operators for image processing. Fuzzy Sets Syst 103:265–275
20. Tizhoosh H (2003) Fast and Robust Fuzzy Edge Detection. In: Nachtegael et al (eds) Fuzzy filters for image processing. Springer, Berlin
21. Liang L, Looney C (2003) Competitive Edge detection. Appl Soft Comput 3:123–137
22. Kim D, Lee W, Kweon I (2004) Automatic edge detection using 3×3 ideal binary pixel patterns and fuzzy-based edge thresholding. Pattern Recogn Lett 25:101–106
23. Russo F (2004) Impulse noise cancellation in image data using a two-output nonlinear filter. Measurement 36(3–4):205–213
24. Yüksel M (2007) Edge detection in noisy images by fuzzy processing. Int J Electron Commun 61:82–89
25. Banerjee M, Kundu MK (2008) Handling of impreciseness in gray level corner detection using fuzzy set theoretic approach. Appl Soft Comput 8(4):1680–1691
26. Várkonyi-Kóczy A (2008) Fuzzy logic supported corner detection. J Intell Fuzzy Syst 19: 41–50
27. Russo F (1999) FIRE operators for image processing. Fuzzy Sets Syst, vol 103, pp 265–275
28. Moravec H (1997) Towards automatic visual obstacle avoidance. In: Proceedings of the 5th international joint conference on artificial intelligence, p 584
29. Schmid Cordelia, Mohr Rrger, Bauckhage Christian (2000) Evaluation of interest point detectors. Int J Comput Vision 37(2):151–172
30. Mokhtarian F, Mohanna F (2006) Performance evaluation of corner detectors using consistency and accuracy measures. Comput Vis Image Underst 102(1):81–94

Printed in the United States
By Bookmasters